Candy Bites

Candy Bites

The Science of Sweets

Richard W. Hartel
AnnaKate Hartel

 Springer

Copernicus Books
An Imprint of Springer Science+Business Media

Richard W. Hartel
Department of Food Science
University of Wisconsin
Madison, WI, USA

AnnaKate Hartel
Marion, IA, USA

ISBN 978-1-4614-9382-2   ISBN 978-1-4614-9383-9 (eBook)
DOI 10.1007/978-1-4614-9383-9
Springer New York Heidelberg Dordrecht London

Library of Congress Control Number: 2014932674

Copernicus Books is a brand of Springer
Springer is part of Springer Science+Business Media (www.springer.com)

We would like to dedicate this book to wife and mother, Paula McMahon. Thanks especially for your patience with us as we wrote this.

Preface

The impetus behind Candy Bites is the candy course taught at the University of Wisconsin-Madison. Every summer since 1963, when the course was initiated in conjunction with the National Confectioner's Association, candy technologists in companies around the world have congregated in Madison for several weeks to learn about candy. From hard candy to chocolate, they learn about ingredients, formulations, and manufacturing methods from experts in the field. They then come down to the candy lab and make numerous batches to understand how formulation and processing conditions influence the quality attributes of each candy.

For example, in the caramel lab, small groups of students make about 15 different caramel varieties. Some use sweetened condensed milk while others use powdered milk, some use butter while others use vegetable fat, and some cook to 238 °F while others cook to 260 °F. At the end, the instructor provides input and evaluation on how and why the observed differences come about.

This long-standing expertise in candy science is also available to the undergraduate Food Science majors at UW-Madison through a senior elective course, Candy Science. From understanding how the boiling point elevation curve influences moisture content in sugar confections, to how the principles of glass transition and the state diagram allow control over candy quality, to controlling the polymorphic crystallization of cocoa butter during tempering of chocolate, the students learn to apply scientific principles to candy making. In this way, candy making becomes more of a science than an art (see Chap. 3).

In this book, we've teamed up to provide a unique product. Most of the chapters were written by Dr. Rich, a Professor of Food Science and lead instructor for both candy courses, with input from AnnaKate, who has degrees in English and Writing. A few chapters were written by AnnaKate, which are indicated as such in the text. It was written so that people with all levels of science education and expertise can enjoy this book. We hope you enjoy learning a little science along with trivia, history, and social insights related to candy

Madison, WI
Marion, IA

Richard W. Hartel
AnnaKate Hartel

Acknowledgments

Many people, too numerous to mention individually, have contributed to this book. Whether reading and editing chapters or providing inspiration for chapters, your contributions have helped improve this work. We thank you all.

Contents

1

Through A Candy Store Window

While on vacation in Provincetown, Massachusetts, a small resort town on the tip of Cape Cod, we took a visit to a local candy shop. The proprietor was making fudge in the window and had a fan blowing the fumes out onto the street as the tourists walked by. Sugar and cream cooked together gives off a wonderful smell reminiscent of caramel and fudge, so many people when they first got a whiff of the exhaust fumes raised their heads to see where the smell was coming from. Nearly everyone, except perhaps for those strict parents who frown on sweets and were intent on getting their kids past a candy store without major incident, looked into the window of the shop to see the candy maker at his fudge kettle.

What a great marketing strategy, blowing your candy smells out onto the street to intrigue the passers-by into coming into your shop. Odor is one of the strongest ties to our deeper emotions and this candy maker was hoping that the smell of cooking fudge would bring out childhood memories and induce people to step into his shop. The fresh candy smell was a better advertisement than a huge banner on the storefront proclaiming a deep discounted sale.

So, of course we went in to look around, see what candies were available, watch the customers searching the shelves for their favorite candies, and to observe the proprietor at his art.

Once inside the shop, we saw that the candy maker was teaching a new employee how to make fudge. It was the start of the summer tourist season and the proprietor was training this young man in the details of making their special brand of fudge. He was showing the employee the proper way to stir as the mixture of sugar, corn syrup, condensed milk and butter cooked in a large

R.W. Hartel and AK. Hartel, *Candy Bites*, DOI 10.1007/978-1-4614-9383-9_1,
© Springer Science+Business Media New York 2014

copper kettle on an open flame. Ensuring that all the ingredients are well-mixed, with the butter properly emulsified, while preventing the milk proteins from scorching on the hot surface requires constant attention and vigorous agitation. The proprietor was teaching the apprentice how to execute a figure-eight mixing pattern with the large wooden paddle to make sure the entire kettle surface was periodically scraped clean without causing a vortex in the middle. I watched with approval since this is exactly the technique we teach in our candy courses.

As the fudge batch cooked on the flame, the proprietor wondered aloud about the hot and humid weather expected for the next few days. "Hmm, it's going to be hot and humid for the next few days" he said. As an experienced candy maker, he knew that the outside conditions could have an impact on the characteristics of his candy—how it would feel and taste, and how long it would last. On a normal day, he would have cooked the batch to a pre-set temperature, defined by the candy thermometer (see Chap. 8), to obtain a smooth, creamy fudge that was firm, but still soft to bite through. The hot, humid weather he knew would make his fudge unacceptably sticky and soft, so he proclaimed to the employee "Let's cook the batch to one or two degrees higher temperature than normal. That'll make it hold better." Without really knowing it, he was applying science to his art.

What was funny was that he then looked over at me, since he knew I was watching, and said "It's not rocket science". I laughed and said, "No, it's candy science." He didn't know who I was. As a scientist (Physics and Engineering) who studies candy making and teaches candy science to anyone who'll come near my lab, I have an appreciation for the science that goes into making a high quality confection (although I often wish I was better at the art of it).

The aroma of fudge cooking on an open flame is wonderfully appealing, so it's not surprising that candy often elicits strong emotions since it's generally tied to childhood experiences. A walk through an old-time candy shop, fudge aroma and all, is often a walk through our childhood. Perhaps for you it's seeing the colored candy dots on the strand of paper or the box of candy cigarettes that brings the memories rushing back. Each one of us has our own buried memories and emotions, just waiting to rush back to mind with the proper stimulus. Candy is one of those stimuli that often create a strong bridge to our childhood memories.

In the following chapters, we hope to build on this image of walking through an old-time candy store to pique your interest about the history, sociology, and especially, the science behind your favorite candies. We hope to provide an entertaining and enjoyable trip back through the candy store memories of your childhood to develop a greater appreciation for the science behind the art of confectionery.

2

All Candy Expo

Like a kid in a candy shop, she flitted from booth to booth. On her left there's a new candy bar to taste (nougat and caramel roll laid on a chocolate wafer) and on her right she's being asked to sample the new Jelly Belly flavors (Dog Food, Dirt, and Centipede). Every way she turns, there's something new and exciting to taste and investigate. It's fun walking through the All Candy Expo, no matter what your age.

The National Confectioners Association (NCA), a corporate sponsored trade group responsible for overseeing the interests of the confectionery industry, holds an enormous annual exposition of all things new in the candy world (now called the Sweets and Snacks Expo). Every year, candy manufacturers and distributors put on their prettiest faces (or hire the prettiest models) to hawk their products to the nation's retailers. Thousands of people come to walk the aisles of the Expo to see what's new.

Unfortunately, not just anyone can attend. Without an invitation, you can't get in. If you own a shop that sells candy, you're invited to this Expo as a buyer. Other than that, everyone else is excluded, except for a candy scientist and his wife. Even the deepest love for candy isn't enough to get you in. You have to be a buyer to go crazy at the All Candy Expo—and there's a good reason for that.

Everyone would love to go crazy in a candy expo!

And many people do. It's almost sad how some people lose it when exposed to such choices. Over the years, the people at NCA have had to change their policies regarding who could attend the Expo and what they could take away because of people's behavior. Children under 16 are no longer allowed in—their behavior,

R.W. Hartel and AK. Hartel, *Candy Bites*, DOI 10.1007/978-1-4614-9383-9_2,
© Springer Science+Business Media New York 2014

goggle-eyed crazy in a candy shop, took away from the intended purpose.

It wasn't just kids, though, that caused problems. Many adults would also go nuts around so much candy. People would roll in luggage carts to fill up with free stuff. Exhibitors often have bowls of candy out for people to sample and these people would completely wipe out the bowl, pouring the contents into their travel bag. How rude—no consideration either for the exhibitor or the next person to come along. Supposedly, one person filled his bag up enough times to fill his station wagon—he took the candy back to his convenience store to sell. Not a bad profit, but at what expense.

To control the greedy nature of people, NCA then limited what bags were allowed into the Expo and, for a while, created a Candy Room to appease people's desires for free candy. Attendees would receive a standard bag as they entered the Candy Room and be allowed to fill it up once. You'd think a free bag of candy would satisfy people, but one bag full wasn't enough for some. Although it was only a few people who found ways to circumvent the rules, for example by building up the walls of the bag with cardboard so it would hold more, NCA finally had enough of people's greed and discontinued the Candy Room.

Expo attendees are now limited to one designated bag to collect samples and brochures and no wheelies allowed—the free candy grab is over.

Too many people just lose it when faced with free candy. They lack control. Or rather, they lose control.

For the most part, we learn to control our urges through the process of growing up. It's not unusual for a young kid to yell "Mine" when another kid tries to play with his toy, but parents generally teach their kids to share and control their selfish urges. Parents also teach their kids not to be gluttons, particularly with sweets and candy. In private, a kid may binge on candy until he gets sick, but at least in public, we grow up being taught to control our inner urges. And those urges for sweets seem to be one of the stronger temptations we face.

But each person is different and we each fall prey to our own temptations. Take, for example, a box of chocolates. Some people can restrict themselves to one piece per night. These people can enjoy the taste of a chocolate and then put the rest of the box aside, knowing it will be there the next night. Seriously, there really are such people—saints. Most of us would go back for another (and maybe even another). Once the taste is in your mouth, it's difficult to stop. Some people have so little control, they'd eat the entire box at one sitting, and then usually regret it.

For what it's worth, if you really want to stop at one chocolate, consider brushing your teeth immediately after that first one. Removing the chocolate taste in your mouth removes the temptation to take another one. Besides, chocolates (and many other things) taste terrible with a toothpaste mouth.

Consider the Marshmallow Experiment. An experimenter and a four-year-old are together in a room. The experimenter says, "You can either have one marshmallow right away or, if you wait 15 minutes, you can have two." He then leaves the room, leaving the four-year-old alone with the marshmallow (and a camera). Imagine the agony. Not surprisingly, some kids succumbed—better one marshmallow now than two later. Others found ways to pretend it wasn't there or had enough self-control to delay gratification for a larger reward. The research found that those children who have the patience to wait are often happier and healthier adults (lower body weight, higher SAT scores, and, in general, significantly more confident) than those who ate the marshmallow right away. The marshmallow test was even a better indicator of future success than socio-economic factors. Marshmallows can tell the future.

As a kid, I'm not sure if I could have waited 15 minutes for a second marshmallow. I was the typical candy fiend, saving money to buy candy, always trying to get the best value for my money (3 Musketeers are really big for their weight and appeared to go a lot farther than the more dense Snickers Bar). Still, it was only a marshmallow; I think I could probably have lasted 15 minutes.

Nowadays, with candy all around me, I can afford to be extremely picky and eat candy sparingly (which, along with lots of cycling miles, helps keep the spare tire at bay as well). People marvel at how I can have candy all around without craving it, but it's probably like anything—you get saturated with it and no longer feel the need all the time.

Still, there's a sense of wonder walking around the All Candy Expo, looking at all the new candies appearing on the market. Although I don't have that same amazement as my wife, there's something really cool about being inside the candy industry and being exposed to all the new sweets.

3

Art or Science: A Brief History of Candy

Numerous articles, blogs and even books have been written about the history of candy. Many of them start with natural sweeteners, like honey and maple syrup, and then move on to refined sugar, which is a relatively modern development. We'll focus here on the history of candy science, since this provides a unique perspective to how candy developed and where we are now.

Is candy-making an art or a science? More than 50 years ago, candy maker Jimmy King of the American Molasses Co. was asked by his peers in the candy industry to give his insight into the difference between art and science in candy making. He suggested that candy making developed over the years as an empirical or "non-rational" art. That is, early candy makers took whatever ingredients were available and experimented with their different attributes until they made something that looked and tasted good. No science was used; it was all trial and error.

Have things changed since then? Candy makers still argue, or at least discuss, this; well at least some candy makers do. In the University of Wisconsin summer candy school, it's not uncommon to hear industry instructors, people with substantial experience in the manufacturing industry, raise the art versus science debate.

Perhaps a brief look at the history of candy development can help shed some light on the art versus science question. The first "candies" were probably fruits and nuts rolled in honey, or something like that. And they were eaten almost immediately, so there were no worries about how long they would last. Not much science in that.

R.W. Hartel and AK. Hartel, *Candy Bites*, DOI 10.1007/978-1-4614-9383-9_3, © Springer Science+Business Media New York 2014

Even centuries later, most candy making was still done without a clue about the science (actually, many things are like this, from food to paint, and even babies). We didn't even know what molecules were until the early 1800s, so how could we understand the details of the candy-making process? Yet our ancestors could still make delicious confections (as they could still make strapping babies without knowing genetics).

The history of candy is intertwined with the development of refined sugar (see Chap. 5). It wasn't until sugar became cheaper and easier to get that candy making really took off. Before that, it was only kings and other wealthy types who could afford to have confections made for them. It's quite probable that the precursors of many of our current candies were developed in king's kitchens around the world.

The 1800s through about the mid-1900s was a time period of intense candy development. Most of our modern candies were either developed or perfected during that time period. While there are numerous new candy introductions each year—the candy industry is continually looking for new ideas—most of the top ten candies have been around for close to 100 years.

Another important development in the candy industry over the past century or so has been technology and automation. Candy used to be made by hand in small batches by artisan candy makers. They had the "feel" of the candy and could often tell when a candy was done by their sensory evaluation (visual, feel, smell, etc.). One of the most amazing tricks used by old-time candy makers was to dip their fingers into the hot cooking mass. They would dip their fingers into cold water, then directly into the boiling sugar syrup (yikes, just thinking about it makes me flinch), and then back out into the cold water, just to tell whether the cooking sugar syrup was ready or not. No science, or thermometer, needed.

Now, most commercial candy is made on large and mostly automated processing lines. Imagine an army of naked Snickers bars on a conveyor passing through a chocolate-fall (a waterfall of melted chocolate). Hundreds of finished candy bars come off the line every minute. Instead of an old-time candy maker to dip his fingers into the syrup to decide if it's done, the most modern technology is used to control every aspect of the operation. This requires a very sophisticated understanding of the science underlying candy manufacture.

Scientifically, as our understanding of the world around us, both macroscopically and microscopically, developed over recent centuries, our understanding of candy making improved as well. From molecules to microbial growth, scientists applied the latest findings to all aspects of our lives, including sweets.

In recent history, the scientific understanding of candy and candy making has grown exponentially. Some of the earliest candy scientists, in the 1940s, 1950s and 1960s, knew an incredible amount about what went on beneath the surface, so to speak. That development continues to this day, with all the latest advances in physics, chemistry, microbiology and even biology being applied to advance our understanding. Many commercial candy companies hire PhDs in a variety of disciplines to help keep them competitive in the modern candy universe.

With this history in mind, is candy-making an art or a science? As with most things, continuous improvements in our understanding of what happens to the ingredients during candy making to make a quality candy is turning candy making into much more of a science-based process.

However, there are still plenty of opportunities for the artistic and creative aspects of confectionery, especially as practiced by artisan candy makers. In fact, one can see a resurgence of artisan candy makers, but perhaps with a difference from past times. They also want to understand the basics of what they're doing in hopes that they can enhance their offerings. Combining the technical knowledge of a science degree with culinary training allows them to develop new and unique offerings.

As one instructor used to say when talking about hard panning (see Chap. 45), it's not an art or a science. . ..—it's a sport. The more you practice, the better you get, whether you understand the science or not. That pretty much sums it up, although as a scientist,

4

Candy Companies Big and Small

A long time ago, confectioners had to do everything themselves in their own shop. Candy making started out as small individually run businesses and, although some candy makers still hold to this tradition, we now find huge international conglomerates dominating the commercial market.

At the beginning, confectioners would make their candy products fresh every day for people to purchase. Down in the alleys of the big cities, along with the signs for blacksmiths, bakers and butchers, would be the shingle for the confectioner, attracting business to his shop. Now, whole towns, like Hershey, PA, have hung out their candy shingle, to attract people from around the world. In Hershey, even the light fixtures are decorated like candy, Kisses to Peanut Butter cups.

One hundred years ago, there were hundreds of candy companies. The late nineteenth and early twentieth century were the heyday of candy development, with many of our national brands developed prior to 1950. Later in the 1900s, however, companies started to grow by buying up others. Like Pac-man gobbling up everything in his way, large candy companies get bigger by gobbling up other candy companies.

Look at the Hershey Company. Known primarily for chocolate, Hershey's has become one of the largest candy companies, primarily by buying out other brands. Twizzlers, Mounds, Almond Joy, York, Kit Kat, Jolly Rancher, PayDay, Zagnut, Zero, Good and Plenty, and the list goes on and on. All brand acquisitions, a business term for buy-out. Mars, Nestle and now the largest candy maker, Ferrara Candy, also have grown by mergers and acquisitions.

R.W. Hartel and AK. Hartel, *Candy Bites*, DOI 10.1007/978-1-4614-9383-9_4,
© Springer Science+Business Media New York 2014

What drives this? The market economy. The key to paying off your stockholders is continual growth. Growth of profits can come about through aggressive marketing, new and bold product initiatives, downsizing and cost efficiencies, or corporate takeovers. Aggressive marketing has always been a cornerstone of the candy industry, particularly with numerous commercials targeting the Saturday morning crowd. As recently as the late 1990s, the candy industry was on a drive to increase per capita consumption of candy. Increasing mindless munching of candy products, everything from Twizzlers to M&Ms, was part of the business mindset. But recent health and nutrition awareness as well as the link to the obesity problem has all but put a cap on marketing, particularly to kids, as a means of business growth. Marketing is still important to brand identity, but the barriers to where and how candy can be marketed continue to rise.

Companies can also grow through new and bold product initiatives and by continually seeking process efficiencies and cost reductions. All companies, not just candy companies, continue to work this approach. Each year, numerous food scientists and engineers are hired by candy companies to find ways to make their products more efficiently and reduce costs. Some also work on the next greatest thing in candy. New technologies often provide unique products for marketers to promote. For example, Caramel-filled Kisses, based on a relatively new technological advance called frozen cone technology, add to the portfolio of Kiss products for marketers to sell.

But the easiest and surest way to ensure continued growth, and continually increasing profit for shareholders, is through brand acquisition. Companies as large as Wrigley are not safe, as the latest acquisition by Mars shows. Even Hershey, one of the big three, was the target of a takeover attempt not too long ago.

All of this conglomeration in the candy business provides a space for new companies to start up. Here's my theory, a candy crystallization theory (loosely) applied to business. Imagine a field covered with individual tents in row after row, lined up so there's no space left for another new tent. Brand acquisition is akin to some tents starting to get bigger and bigger by incorporating all the tents around them. Due to the efficiencies that come from being larger, when two tents incorporate together, the larger tent doesn't need to take up the same space as two separate tents. As tents get bigger, more and more space opens up between the remaining tents. From a crystallization standpoint, the open space formed as big tents get bigger and bigger allows new tents to "nucleate" in the open space.

So, as candy companies get larger and larger, a business space opens up that allows new and creative ideas to develop a market. It may be a new and innovative product, it may be a high-quality niche, or it may be something else that's really outside the tent, but the business climate is ripe for new market additions.

Although artisanal, local chocolatiers certainly operate in the space between the big tents, we're talking about what might be considered candy companies, with unique candy products, not truffles. A couple examples will show what we mean.

A really good example is a company called Unreal Candy. Is there a way to make candy healthy? Isn't that a great idea in these days of health awareness? Unreal has tried by reinventing some popular favorites, like Reese's Peanut Butter Cups and Snickers, "from 100 percent REAL ingredients. Real milk chocolate, real caramel, real nougat, real peanut butter and real cane sugar. No artificial stuff, no corn syrup, no hydrogenated oil, no preservatives, no GMOs, and less sugar." As they say in their marketing, Unreal Candy has taken out the "bad" stuff and left in the "good" stuff. As they put it, they've "Unjunked®" several popular candies by taking out the ingredients that deter some people from eating the commercial stuff.

Another company that's creating new candy bars is Zingerman's Candy Manufactory. With candy bars like Zzang! Original, Ca$hew Cow, Wowza, and What the Fudge?, they're trying to find a spot outside the big tents to sell these new candy bars. These are not knock-offs of current commercial products, these are primarily new and intriguing candy bars. Wowza is made of raspberry chocolate ganache, raspberry nougat and raspberry jellied candies, all coated in dark chocolate. The Ca$hew Cow contains "freshly roasted cashews and cashew brittle with milk chocolate gianduja enrobed in dark chocolate." What's gianduja? Even we had to look it up. It's chocolate and hazelnut paste, a fancy name for Nutella, but that's part of the marketing to distinguish their product from others.

Who will be the next Milton Hershey or Forrest Mars? Who knows, but Jack, a budding young candy maker who visited us during our candy course a few years ago, says his goal is to become the largest candy maker in the world. I joked that he better be; you should never trust a skinny candy maker, right? Fat candy jokes aside, with the current market for takeovers and corporate consolidation, I think there will be plenty of space for Jack to pitch a tent and grow his business.

5

Sugar History and Production

Buy American. Only it's getting harder and harder to buy American-made candy. Imagine competing in a global economy when we penalize American companies by making them pay nearly double the price of the world market. Many of the larger candy companies have shut the doors to their American plants and built new facilities in Canada and Mexico, where they can buy sugar at world prices.

What's behind this? The Sugar Act. Enacted during the Great Depression, the intent of the Sugar Act was to protect American sugar farmers by restricting imports and providing subsidies to support crop yields. It's done that, but the Sugar Act has really helped Big Sugar—the Sugar Barons, the families that own the sugar cane plantations.

Well, it's actually way more complicated than that, as political dealings usually are. The bottom line is that noble efforts to protect the American sugar industry have reached the point where many candy companies can't compete.

Let's look at a brief history of sugar and how we got to this point.

As with many things, foods in particular, it's hard to say exactly when our ancestors recognized sugar as something valuable. According to various accounts, there is archeological evidence that sugarcane was first developed as a crop in New Guinea around 8,000 BC. It slowly spread throughout Southeast Asia and into India. At first, people probably simply chewed the cane for the sweetness; perhaps some people even extracted the juice to drink. Somewhere in early to mid first century, people in India developed a method of crystallizing the sugar in the extracted juice. Most likely, they left the juice out in the heat one day, the water evaporated off and crystals accidentally formed. What was probably a serendipitous discovery ended up being a major step in the development of sugar and candy.

Sugar crystals are far more stable than the juice. The extracted juice is fairly dilute, with only a few percent of sugar. It's quite prone to microbial growth, especially in the warmth of the tropical climates where sugarcane grows. Unless you're making rum (another off-shoot of the sugarcane industry), fermentation of the cane juice is a problem. Learning how to form stable sugar crystals probably led to it being spread over farther distances, making its way into China and through the Arabian peninsula and eventually on into Europe.

The Greeks and Romans knew of sugar, but not as a food commodity. In Roman times, sugar had a reputation as being medicinal, providing relief for gastrointestinal problems. It was during the "Arab agricultural revolution" in Medieval times that sugarcane spread widely through Mesopotamia, with larger-scale processing becoming common. The Crusaders then brought sugar back with them to Europe. Once Europeans developed a sweet tooth, sugarcane agriculture blossomed to meet the demand.

To feed the growing sweet tooth, Europeans started looking for places to grow sugarcane so they could control the industry. The New World was the most likely place; Columbus supposedly carried some sugarcane plants on his second voyage to the Americas. The Portuguese carried it to Brazil and the Dutch to the Caribbean. A growing number of sugar mills were already producing sugar in Cuba and Jamaica by the early 1500s.

Technology really advanced in these years as well. The process of refining sugar involves several steps. First the juice is collected through a sequence of crushing the cane followed by hot water extraction. Water extracts all sorts of compounds from the cane, with sucrose only present at a few percent. To improve yields, the juice must be clarified to remove impurities. This clarified juice is then concentrated by evaporating off the water. When the juice becomes supersaturated, sugar crystals are formed, which can be separated from the liquid and dried to form raw sugar. Further refining operations involve recrystallization of the sugar several times to create the pure white crystalline powder that we now know as refined table sugar. The modern process has been automated and upgraded to be orders of magnitude more efficient than in the early days.

Since growing sugarcane and processing it into refined sugar, particularly in the early days, is extremely labor intensive, a cheap source of labor was needed. Unfortunately, that source of cheap labor was primarily slaves from Africa. Boats brought slaves from Africa to the New World and returned the sugar to Europe to fill the growing need. By the early eighteenth century, sugar was widely used in Europe, initially for sweetening tea but eventually it was also turned into sweet treats like candy.

During the Napoleonic Wars, trade embargoes threatened the sugar supply in Europe. The sugar beet, discovered in 1747, became a viable alternative. Because the sugar beet is grown in moderate climates, rather than in the tropical climates required for sugarcane, Europe now had a source of sugar that was not dependent on importing from other countries. Although sugarcane still dominates, the sugar beet industry today supplies about 30 percent of the sugar consumed.

The United States produces both sugarcane and sugar beet. Sugarcane is predominantly grown in Florida and Louisiana, while the upper Midwest supports sugar beet farming. Although the United States is not one of the top producers of refined sugar, the industry remains a viable one, and one that's protected by the government as an important commodity. And that's where today's problems arise.

Is the current situation caused by the Sugar Barons and their lobbying to maintain the status quo, as some think? Of course not, the problems are much deeper than that. Even though the Sugar Act helps protect farmers, the sugar barons appear to have gotten a sweet deal from government control.

Regardless of what's led to the current situation, reform is needed to bring balance to the situation. Asking candy manufacturers to pay well above the world price is causing them to take their plants and jobs to other countries, to the deterioration of the local economy. If things continue like this, it will be even more difficult to find American-made candies and the ones that are available will be significantly more expensive than they are now.

6

The Demon Sugar

How fast things have changed over the years. The past century or so has seen an amazing rate of change in almost every aspect of humanity. In transportation, horse-drawn carriages have given way to cars with remote sensing to protect us from ourselves; in lighting, we've gone from candles to laser beams; and in computing, from the abacus to supercomputer phones that fit on your wrist (calling Dick Tracy). Almost all facets of our lives have changed tremendously. Our perception of food, especially candy, has evolved considerably over that time as well.

In the earliest days of sugar, it was a status symbol—teeth blackened by sugar were considered a sign of wealth in Elizabethan days. Later, as sugar became more available and new candies were being developed, it was seen as a splendid treat and a source of needed calories, even valued nutrition. Fifty years ago, Kraft caramels were touted as not just being delicious but nutritious too. One old label said "proteins and minerals of 20 ounces of milk in every pound," as if it was better to eat a pound of caramels than drink a large glass of milk (although milk has its detractors too). Fast forward to today and sugar, and candy by association, is considered by some to be a toxic poison that causes nearly all of mankind's ills.

The current arguments against sugar are numerous—one web site actually quotes 143 reasons why sugar ruins your health (with a side bar that "Sugar Kills!"). From contributing to juvenile delinquency, reducing learning capacity and leading to alcoholism (from liqueur-based candies?), this list appears to blame almost every health-related condition on sugar (and by association, on candy). Heck, sugar is so bad that they claim it even ruins a person's sex life.

R.W. Hartel and AK. Hartel, *Candy Bites*, DOI 10.1007/978-1-4614-9383-9_6,
© Springer Science+Business Media New York 2014

Sensationalism aside, there is clear evidence that sugar can have negative effects on our health, if misused. For one, it can be a contributor to metabolic syndrome, the complex chain of events that lead to diabetes and cardiovascular disease. It's primarily "empty" calories, so it at least contributes to obesity and the attendant health woes that go with that, though as with all such issues the story isn't so simple—sugar isn't the only cause. It also contributes to adult-onset diabetes through modulation of the insulin response. It can cause cavities. Some claims have been made that it might be addictive in humans (although this hasn't been proven). And it causes good kids to go all hyper, the so-called sugar high.

Wait, is there really such a thing as a sugar high? Depends on what you mean by that. If it's kids going hyper after eating sweets, then the answer is a pretty definitive no, despite the fact that blood glucose levels can change dramatically after you eat a candy bar. Numerous studies have fed kids, and adults too, either sugar or placebo and evaluated behavior. The data clearly show that neither kids nor adults exhibited any evidence of hyperactivity with either treatment. Instead the sugar high myth is often attributed to "confirmation bias", where observations confirm beliefs. We often see behavior we think we expect to see; mothers who thought their kids got a sugar high were more likely to say their kids were hyperactive even when given the placebo than mothers less inclined to believe there is such a thing as a sugar high. Apparently, the excitement of the occasion for eating sugar (birthday party, Halloween, etc.) is the most likely cause of kids going hyper.

There is certainly an increase in blood sugar after eating sugar-rich foods, the glycemic response. But glycemic index by itself is not the whole story, since the effects of pure sugars on blood glucose is less than that of more complex carbohydrates like bread and pasta. This glycemic response triggers a release of insulin, the body's hormone for utilizing and storing glucose in the blood stream. It's when the body's response systems get out of whack that problems occur, as in metabolic syndrome, and excessive sugar consumption can contribute to that.

Sugar is getting such a bad rap these days, in fact, that even fruit has come under suspicion. Fruits are high in sugars, mostly glucose and fructose, the simple sugars that cause a significant glycemic response. Even though whole fruit provides valuable fiber, antioxidants and other valuable micronutrients, some people are worried more about the sugar response than the healthful effects from eating fruit. In fact, recent studies clearly show that the fiber in fruits actually moderates the glycemic and insulin responses to fruit consumption. So please, eat all the fruit you can.

It's also well known that it's far better to eat a whole fruit, including the peel where appropriate, than to drink the clarified juice. Most of the "healthy" components of the fruit are in the fiber and insoluble solids. For example, the total polyphenolic compounds in clarified juice are more than fivefold less than in the intact fruit, including the peel. Even cloudy juice, or apple cider, had only slightly over half of the polyphenols. Eat fruits often, as intact as possible.

Sugar can have other effects on humanity as well. One clear case of sugar being bad occurred in 2008 at a sugar refinery in Georgia. Refined sugar in crystalline powder form is actually quite explosive. If a layer of fine particles forms on a hot surface and there's a spark, it's possible that a powder explosion will ensue. In the Georgia sugar refinery, the five elements required for a powder explosion came together with a disastrous result: sugar was the (1) fuel in the form of a (2) fine powder dispersion, (3) confined to a small space with a spark for (4) ignition with plenty of (5) oxygen available. The explosion and subsequent fire effectively leveled the plant.

Then, there was the Great Molasses Flood in Boston in 1919. A huge tank of molasses ruptured. Over two million gallons of the thick stuff was released, generating a molasses street-tsunami that destroyed almost everything in its path, including buildings and trucks. It even damaged the girders holding up the elevated train. Both people and animals died, and many more injured.

So, is sugar really that bad for us? Besides the sugar catastrophes noted above, there are very real effects of sugar on our body. Eat too much of it, especially the refined sugars, and yes, problems ensue.

7

The Sweet Tooth

Imagine hearing a musical tone and sensing sweetness. Such is the life of one recently diagnosed synesthete who crossed senses of hearing and taste. Synesthesia is a neurological condition where stimulation of one sense leads to an involuntary response in another sense. Most synesthetes associate music with color. It's quite rare that someone associates a musical stimulation with a sensory response.

Most of us need to put sugar, or some other sweet substance, into our mouth to sense sweetness. This sweetness, of course, is the basis of why we enjoy candy so much. In fact, enjoyment of the sweet taste is apparently inherent in humans. Even new-born babies smile when a sweet solution is dropped on their tongue (and grimace from a bitter solution).

How did we come to learn which compounds are sweet and which are not? Trial and error. We humans are infatuated with putting things in our mouth—just watch any toddler exploring his environment. Everything he touches goes right into his mouth as if that's the only sense that works at that age. I imagine the same thing with our ancestors as everything from tree sap to grass went into their mouths. Eventually, humankind developed a compilation of things that were sweet.

There are a number of compounds that give a sweet sensation although not all are appropriate for candies or even to put in your mouth. The usual sugars top the list of sweeteners in confections—common compounds such as sucrose, glucose, and fructose are usually the top ingredients in candy. Corn syrup, a mixture of glucose and its polymers derived from corn starch, provides some

R.W. Hartel and AK. Hartel, *Candy Bites*, DOI 10.1007/978-1-4614-9383-9_7,
© Springer Science+Business Media New York 2014

sweetness because of the glucose and maltose present. Note that in other countries, wheat or potato are the source for these starch-based syrups, collectively known as glucose syrups.

Not each of these sugars provides the same level of sweetness. Sucrose is always used as the standard, with 100 percent on the sweetness score. Glucose is only about 60 percent the sweetness of sucrose, whereas fructose is significantly sweeter, upwards of 175 percent. This is why regular corn syrup (glucose polymers) is less sweet than sucrose but high fructose corn syrup is a little sweeter. The actual sweetness ratio depends to some extent on exactly how the comparison is made, so the numbers usually vary slightly. Lactose, the natural sweetener found in milk, is actually a lot less sweet compared to sugar, at about 15 percent. But even at this low level, lactose gives milk its slight sweetness.

The sugar alcohols used in sugar-free candy are generally less sweet than sucrose, with the exception of xylitol. Xylitol is essentially the same sweetness as sucrose, but other common polyols, like sorbitol and maltitol, are less sweet, about 50 and 90 percent, respectively. Isomalt, a sugar alcohol derived from sucrose and commonly used in sugar-free hard candies, is only about 40 percent as sweet as sucrose. That's why sugar-free gum and candies often contain high-intensity sweeteners.

High-intensity sweeteners may be defined as compounds that are significantly sweeter than sucrose. In fact, some are so sweet that only a single speck will overwhelm your sense buds. Supposedly the sweetest compound is one called lugduname. At about 220,000–300,000 times as sweet as sucrose, it doesn't take much. Since it's not approved for food use, you don't need to worry about sweet shock from lugduname.

But there are plenty other high-intensity sweeteners available for use. Common sweeteners (with their approximate relative sweetness) include aspartame (200), acesulfame-potassium (200), saccharin (300), and sucralose (600). Some proteins are significantly sweeter than sucrose—these include allitame (2,000) and thaumatin (3,000). Recently, an extract from the stevia plant has been promoted as a natural sweetener. Stevia is actually a mixture of different steviosides and rebaudiosides that have varying sweetness, up to about 300 times that of sucrose. Because of its stability and natural origin, stevia appears set to sweeten our lives much more in the future.

High-intensity sweeteners are also often called low-calorie sweeteners because you need far less of them to provide the desired level of sweetness. This is good in soft drinks. Diet sodas are essentially flavored water with a low level of high-intensity sweetener instead of sugar, so have few calories. Effectively, the sugar has been replaced with water. This approach doesn't work as well in candy.

If you replace the sugar in candy with the equivalent sweetness of sucralose, say, you would have only one six hundredth of the mass. Sugar in candy also provides bulk and that needs to be replaced when high-intensity sweeteners are used. These bulking agents (polyols, polydextrose, inulin, among others) also carry calories so the calorie reduction in candies is not as significant compared to soda.

What is it that causes a compound to promote a sweet taste in our mouth? The chemical make-up of sweet compounds is extremely diverse, from proteins to simple carbohydrates, and it's not clear why they all give similar sweet taste. This is actually a very active area of research as we uncover more about our taste buds and how they cause a response in the brain; the theories have changed in recent years as our understanding grows.

Earlier theories of sweetness were related to the geometry of the molecule and how it interacted with the taste buds on our tongue. In one common theory, a sweet molecule needed a hydrogen donor within 0.3 nanometers of a Lewis base in order to associate with the sweet taste bud.

Recent studies are beginning to show the biochemical connection between the sweet molecules released in your mouth and the perception of a sweet taste. A complex set of events occurs, initiated by the chemical association between the molecule and the sensors of your taste buds. There are actually two different sensors on your "sweet" taste buds that respond to sweetener molecules, and each allows interaction with molecules of different chemical structures (sugars, proteins, etc.). This is probably why so many different molecules exhibit sweetness, because your taste buds were designed to accommodate these differences and still provide that sweet taste.

Once the sweetener molecule has docked on the sensors, neurotransmitters are released to the brain. The brain then processes that signal in the context of any other pertinent sensory information—the ultimate result is the taste of "sweet". The sense of sweetness may be innate, as seen by the smiling baby example, but it's also tempered by our experiences and the situation.

The search for sweetness without the calories has gone into overdrive in recent years. Perhaps spurred by the obesity problem, scientists all over the world are virtually putting all sorts of things in their mouths to test for sweetness. This time around, however, we're guided by our increasing understanding of the physiology of sweetness. It's no longer completely trial and error.

What's the future of satisfying humankind's sweet tooth without causing other problems? Perhaps the candy of the future will induce sweetness through synesthesia, rewiring our brains to taste sweet when we hear music. That would be an interesting way to satisfy a sweet tooth without the calories.

8

Soft Ball to Hard Crack

If you've ever made candy at home, you've probably used a candy thermometer to tell you when you reached the right stage in cooking sugar syrups. Not that you really need a thermometer—old time candy makers could tell how their syrups were doing by look and feel. Yeah, feel. Stories of candy makers who would dip their fingers into hot boiling sugar syrup are not exaggerated. We've seen them.

The old time candy maker would dip his (yes, most were men) finger into the boiling syrup and quickly dip them into a cup of cold water. The trick to not burning your fingers was to cool them off first by dipping the finger into the cold water and then dipping it into the boiling syrup. Done correctly, no damage is done. Done wrong; well, I hear the new cures for third degree burns are pretty amazing.

Better to use a thermometer to see how your candy is doing.

A candy thermometer isn't really different from any other thermometer—it still reads the temperature of whatever you stick it into. What's different is that the candy thermometer has some important candy "benchmarks" etched into the base alongside the actual temperature value.

At 230 °F, the candy thermometer says you've reach the thread state, but cook your sugar syrup to 305 °F and you're at the hard crack state. Soft ball (235 °F), firm ball (245 °F), hard ball (260 °F), and soft crack (280 °F) are all milestones that fall between. These marks correspond to the types of candy that can be made.

What's the difference between soft ball and hard crack? Candy makers came up with these terms because they describe exactly

R.W. Hartel and AK. Hartel, *Candy Bites*, DOI 10.1007/978-1-4614-9383-9_8, © Springer Science+Business Media New York 2014

what state the candy syrup takes on when it reaches those temperatures. To understand these states, we need to first talk about boiling sugar syrups. In our candy class, we spend lots of time watching sugar syrups boil—it's akin to watching paint dry or corn grow. But there's some interesting science going on.

In school, you're taught that water boils at 212 °F and its temperature stays right there (at 212 °F) for the entire time it's boiling—until the very last drop of water is turned into steam. That's not true with sugar syrups because the sugar chemically interacts with the water to change its boiling point. Sugar molecules make it more difficult for the water molecules to evaporate off the surface to make water vapor. Since water boils when the pressure of its vapor reaches atmospheric pressure, the fact that sugar reduces that vapor pressure (evaporation from the surface) means that it raises the boiling temperature. With sugars, you have to raise the temperature higher to get the vapor pressure to reach atmospheric pressure.

So if a little sugar makes the boiling point of water go up a little, then a lot of sugar should make the boiling point go up a lot, right? Right. In fact, the amount that the boiling temperature goes up is proportional to how much sugar is there—the higher the concentration of sugar, the higher the boiling temperature. So when water is boiled off from a sugar syrup, the concentration of sugar goes up—which means its boiling point goes up and the concentration goes up further. And so on.

As boiling temperature goes up, the sugar concentration goes up, and so does the viscosity of the sugar syrup. And it's that viscosity that gives rise to the terms on the candy thermometer. Those terms represent the viscosity, or physical state, of the sugar syrup when the candy maker drops some of the boiling sugar syrup into cold water. Some candy makers would use their fingers—we don't recommend that since hot sugar syrups, especially those at 300 °F, are so hot and viscous that if you get some on you, you can't make it to the sink to rinse it off fast enough to avoid third degree burns. Many candy makers have their own stories, and usually scars, to prove this. Here, look at this finger.

Let's go back to look at viscosity of the sugar syrups as they're cooked. At the soft ball stage, at about 235 °F, the syrup forms threads in the cold water that can be gathered into a ball between the fingers—sort of. As a soft ball, the sugar mass is still soft and flows between the fingers. It is not viscous enough to hold its shape. Candy made from this syrup will be soft and runny. Soft, gooey caramels, good for filling into chocolates, are cooked to this temperature.

At 244 °F, enough water has boiled off that viscosity is significantly increased. When plunged into cold water, sugar syrup boiled to 244 °F forms a firm ball. It can be deformed easily, but still holds its shape and doesn't flow under the force of gravity (a phenomenon called cold flow). Chewy caramels fit in this category.

By the time the candy thermometer reads 260 °F, the concentration of the sugar syrup has gone up substantially, as has its viscosity. After being dropped into cold water, the sugar syrup can still be formed into a ball, but this one is hard and retains the ball shape. It is sufficiently viscous that the sugar syrup stands up to its own weight for a very long time. Salt water taffy and really chewy caramel fit into this candy category.

Sugar syrup cooked to 300 °F and cooled quickly in cold water forms hard, brittle threads that crack when you snap them—thus, the hard crack state. In fact, sugar cooked to this temperature and cooled quickly to room temperature turns into a sugar glass—an amorphous matrix of sugar molecules that has solid-like characteristics. Hard candy and brittles are cooked to 300 °F to form sugar glasses.

In fact, the temperature measured by a candy thermometer (or any thermometer for that matter) actually becomes a measure of the sugar concentration, or by difference, the water content. And water content is one of the key elements that the candy maker needs to control. Measuring temperature is a snap compared to measuring water content, especially in very viscous candies. So the candy thermometer provides a quality control tool to the candy maker to obtain exactly the right viscosity for the product.

9

Breakaway Glass: A Soft Solid

If you were hit over the head with a glass bottle or thrown through a pane of window glass, it would hurt and be dangerous. That makes the stunt actors in the movies (remember those saloon scenes in the old Westerns) either crazy or well paid, or both. Either that or they know something we don't—they don't use real glass. Wait, yes they do—it's just not glass made from the same materials that make normal window glass.

Window glass is made primarily from silica, with other additives to moderate it's properties, and when shattered, it has sharp edges that can cause serious damage. That's why your parents told you not to play with glass—it's dangerous. But the class of materials called glass, of which window glass is only one example, extends well beyond the common silica glass of windowpanes.

One such material is sugar glass, the base for such candies as Jolly Ranchers and LifeSavers, and also lollipops, root beer barrels, lemon drops, Pop-Rocks and many more. Even cotton candy is a sugar glass (see Chap. 10), although it behaves more like fiberglass than a windowpane.

Sugar glasses are made by cooking sugar syrups to temperatures in excess of 300 °F to boil off water. Enough water is driven off that there's only about 2 percent water left. When cooled quickly by dropping the hot syrup into cold water, the sugar takes on the "hard crack" state (see Chap. 8). The sugar syrup solidifies instantaneously into strands upon entering the cold water and when the strands are removed from the water, they're brittle and easily cracked—hence, hard crack.

R.W. Hartel and AK. Hartel, *Candy Bites*, DOI 10.1007/978-1-4614-9383-9_9,
© Springer Science+Business Media New York 2014

But is a sugar glass a solid? What exactly is a solid? The dictionary actually provides all sorts of definitions. One can own solid gold, be a solid citizen, be financially solid, and give a solid performance, none of which help here. Scientifically, a solid is defined as something "of definite shape and volume; not liquid or gaseous." Wait, a solid is something that's not a liquid or a gas? Didn't your grade school teacher pound into your head that you don't define something by what it's not? Yes, there's some ambiguity with the term solid.

Some things we consider to be solid include crystalline materials like salt and ice, and noncrystalline materials like wood and window glass. Some materials can form either crystalline or noncrystalline solids, depending on how they're processed. Examples include metals, rocks and sugar. In confections, solid sugar can be either a glass, as in cotton candy (see Chap. 10), or a crystal, as in rock candy (see Chap. 11).

Physicists don't consider a glass to be a true solid; it's just a viscous liquid. Glass is often called an amorphous solid, to distinguish it from a crystalline solid. In an amorphous solid, the molecules are randomly "frozen" into space to form a glass, as opposed to the uniform orientation of molecules in a crystal lattice. Another way to think about it is in terms of molecular ordering—the molecules in crystals have long-range order (a repeating pattern), whereas molecules in glasses have only short-range order (essentially random).

A glass is solid enough that it hardly flows, holding its shape for years; although technically, as a liquid, it will flow given sufficient time. Consider the informative pitch drop experiment. In 1927, scientists at University of Queensland in Australia set up a rig with some "solid" tar in an open funnel and waited to see what would happen. And waited and waited. Over the past 90 years, only about eight drops have fallen from the original pitch ball. Despite the paucity of drops, this experiment clearly demonstrates how something so solid-like can still flow. For their efforts, the two scientists were awarded an "Ig Nobel" award in 2005. This award, sponsored by the journal Annals of Improbable Research, honors "achievements that first make people laugh and then make them think."

People often cite old European cathedral windows as another example of how a glass flows over a long period of time. It turns out that many cathedral windows are thicker now at the bottom than at the top, suggesting that they flowed over time. If true, this would be a perfect example of the Deborah number. Named for the biblical quote that "mountains flowed before the Lord", the Deborah number is, loosely, the ratio of how long something takes to deform to how long we have to watch it. At low Deborah numbers, materials are fluid-like (deform quickly so we can see it right away) while at high Deborah numbers, deformation is so slow that we need the Lord's perspective, a very long time, to observe flow. Others argue, however, that glassmaking hundreds of years ago was not perfect and these variations in thickness may have been inherent in the glass blowing process. We'll probably never know for sure.

These considerations have led to a whole new field, called soft matter physics. Although soft solids span a wide range of materials and industries, from window glass and cosmetics to rubber tires and cardboard, numerous foods can be considered soft solids. Bread, Cheerios, yogurt, Cool Whip, and most other "solid" foods, are soft solids. Arguably one of the clearest examples of a soft solid is Jell-O salad, jiggly mounds of Jell-O with fruit and sometimes carrot suspended throughout. Hard candy, a sugar glass, is also a soft solid, albeit significantly harder than Jell-O.

In old Western movies, breakaway glass used for windowpanes through which bad guys were thrown and whiskey bottles to hit bad guys on the head were made from sugar. Sugar glass is a lot softer and less brittle than window glass. Shards of breakaway sugar glass don't hurt like shards of real window glass, yet sugar glass has exactly the same appearance as window glass. When made correctly, sugar glass can be as clear and transparent as window glass. Hence, its use in the old Westerns.

Breakaway glass for movies and plays is no longer made from sugar glass—better materials have been developed. Any candy maker knows that sugar glass is extremely sensitive to heat and moisture. Scenes in old Westerns where bad guys were thrown through windows had to be shot in the early morning, shortly after the breakaway sugar glass was made. As soon as the day warmed up, the sugar glass started to get sticky; if left too long, eventually it would flow.

Modern breakaway glass is made from materials other than sugar. For example, one type of breakaway glass is made from a urethane liquid plastic. Just mix the two liquid components together, pour into the mold and allow to set. When solidified, it's transparent and shatters in the same way as window glass, yet it has no sharp edges to cut through skin.

Breakaway glass is a great example of how soft matter physics improves our lives. Now, Western stuntmen no longer have to worry about getting cut when getting bopped on the head with a glass bottle or being thrown through a glass window. Well, as long as the glass bottle is only an eighth inch thick. Even "soft" solids can hurt when you get hit over the head with them.

10

Cotton Candy

Summer time brings county and state fairs, and the foods we associate with them. Along with hot dogs and deep-fried dough, another carnival favorite is cotton candy. Most kids are drawn to the machine that spews out flavorful strands of candy collected on a paper cone.

Sometimes called candy floss or fairy floss, cotton candy probably originates from a product we now call spun sugar. Spun sugar is made by heating sugar syrup to the hard crack stage (see Chap. 8), covering a fork (or a whisk with the end cut off) with the molten sugar, and then allowing threads of sugar syrup to fall off the tines across a roller or spun over a bowl. The thin strands of essentially pure solidified sugar can be formed into various shapes to make intriguing and sweet desserts. When the molten sugar solidifies, the molecules are "frozen" into space. Lacking sufficient mobility to organize into a crystal, they remain randomly oriented in an amorphous glass.

Although several inventors lay claim to being the first to develop a machine for making sugar floss, apparently the first people to receive a patent for such a process were the inventors Wharton and Morrison. In 1899, they figured out a way to force molten sugar through holes in a screen to form a floss of solidified sugar, which was then collected on a paper cone to make a treat. They sold fairy floss at the 1904 St. Louis World's Fair for twenty-five cents a box.

To make cotton candy these days, flavored granulated sugar is poured into the center of the "spinner", a spinning disk with small holes in the edges. The heating element, similar to that found in a

R.W. Hartel and AK. Hartel, *Candy Bites*, DOI 10.1007/978-1-4614-9383-9_10, © Springer Science+Business Media New York 2014

toaster, is turned on to melt the sugar. When the sugar crystals melt, the liquid sugar melt streams out of the tiny holes on the outside of the spinning disk. As soon as the thin stream of liquid sugar (or floss) hits the colder air, it solidifies into a sugar glass, called cotton candy. Collect the candy floss on a paper cone and you have a yummy fair treat.

While cotton candy vendors at the fair use electric motors to run the heater and spinner, we found an example of a more environmental approach. At a market in China, one enterprising candy seller rigged a bicycle to make cotton candy. Instead of an electric heater, he used a propane torch to heat the sugar crystals, with the spinning device being powered by the bicycle wheel. I wonder how many miles of biking it took to spin a cone of cotton candy?

From a molecular arrangement standpoint, cotton candy (sugar) has a lot in common with window glass (silica) and breakaway glass (see Chap. 9). The red-hot bulb of fluid silica on the end of the glass-blower's pipe is a lot like the molten sugar inside the cotton candy spinner. In both cases, the molten fluid can be formed and shaped, but when cooled, it solidifies into a glass in whatever new shape is desired. Dessert chefs work with spun sugar glass to make fancy sculptures in the same way that glass-blowers make fine "crystal" art pieces.

Have you ever wondered why fancy glass art pieces are called "crystal" glass? As with jumbo shrimp and nondairy creamer, the expression "crystal glass" is an oxymoron, since a glass can't be a crystal. In a crystal, the molecules are organized into a uniform pattern, or crystal lattice. In contrast, the molecules of a glass are just randomly oriented. In fact, fancy (and expensive) crystal glass is distinguished from regular glass by its composition. The presence of lead oxide (or now, barium, zinc or potassium oxide because lead is a hazard) gives silica glass unique decorative properties, and allows it to be called crystal glass. But molecularly, it's a glass, not a crystal.

Interestingly, fiberglass and cotton candy also have a lot in common. Fiberglass, first commercialized by the Owen-Corning Fiberglass Corporation in 1938, is made by extruding molten silica glass through small holes to make thin strands or fibers of glass. As the strands exit the extruder, they cool into the solid glassy state and are collected for further processing. The process is essentially the same as for making cotton candy.

One of the things that distinguish cotton candy from fiberglass is sensitivity to heat and water. Any humidity or excessive heat and the sugar glass collapses. Imagine a fine spun-sugar sculpture sitting out on a hot, humid day. It would get sticky, and start to flow pretty quickly. Luckily, fiberglass doesn't have this problem so it works well as insulation material. Moisture is also why you have to eat your cotton candy cone pretty quickly. The cotton candy picks up moisture from the humid air, allowing the sugar molecules sufficient mobility that they crystallize and the cone collapses.

How then can you buy cotton candy at the store if it's so susceptible to moisture uptake and collapse? The packaging. As long as the cotton candy is contained within an absolute water barrier, like a foil package that is well sealed, no moisture can get in to cause collapse. The candy will last for years without change. This is an example where the package cost far exceeds the cost of the product within. The cotton candy within the package probably cost just a few cents to make.

Interestingly, where one person sees a problem, another person sometimes sees an opportunity. One inventor used the water-loving properties of cotton candy to his advantage in developing an aspirin pill that doesn't need to be swallowed. First, he made cotton candy floss that contained the right dose of salicylic acid (the chemical compound in aspirin). That floss was put into a press and squeezed together to make a tablet, with the individual floss strands coming together from the pressure to form a tablet. Pop it into your mouth and, because the sugar glass likes water so much, the tablet dissolves on your tongue before you can swallow. Don't like swallowing pills? Take your aspirin in a quick-dissolving tablet; so quick that it dissolves even before you have a chance to swallow.

When is National Cotton Candy Day? Numerous sites on the internet say December 7 (a "day of infamy"). This was confirmed at the Russell Stover site. One site said November 7. Several sites say

July 31. One site even says it's both July 31 and December 7. Why all the confusion about this? Which is correct? Based on the number of citations, December 7 wins hands down. However, we'll celebrate cotton candy on June 11, the date listed at the National Confectioners Association website. We figure the national candy organization would know best.

11

Rock Candy

Big Rock Candy Mountain. Whether you're a fan of the original lyrics of Harry McClintock, the cleaner version of Burl Ives, or the modern kids version, the message of Big Rock Candy Mountain is still the same—life is sweeter there. Whether you prefer bees buzzing in cigarette trees or peppermint trees, Big Rock Candy Mountain is a fine place to be.

What is rock candy? Essentially it's just large sucrose crystals, either stuck to a stick, hanging on a string, or as loose, individual crystals. It's really easy to make, but it takes a long time.

Rock candy was supposedly discovered in China many years ago by accident. As the story goes, a young woman was sneaking a bowl of sugar syrup when her boss came in. To avoid getting caught, she quickly poured her bowl of syrup into a can of lard, covered it with bran and hid the can in the firewood. Days later she retrieved the can and found the sweet crystals.

Everyone has made rock candy, right? If you haven't, here's the process. Heat up some water in a pot on the stove, dump in some sugar to get it to dissolve, then take it off the heat, pour it into a jar and allow it to cool. To help nucleate sugar crystals, it helps to use a rough stick or a string, or even a paper clip would work. Insert into the jar of sugar syrup and wait. For days.

Making rock candy is based on two principles—generating a supersaturated solution, which is based on the temperature dependence of the solubility of sugar in water, and crystallization, specifically the processes called nucleation and growth.

Solubility can best be demonstrated over a cup of tea. Would you like yours hot or cold? Pour two packets of table sugar into iced

R.W. Hartel and AK. Hartel, *Candy Bites*, DOI 10.1007/978-1-4614-9383-9_11,

tea and then do the same for a cup of hot tea. Stir and all the sugar dissolves in the hot version while myriad undissolved crystals remain at the bottom of the iced tea. That's because the amount of sugar that can dissolve in water depends on the temperature—higher temperature, more sugar dissolves.

Many materials are soluble in water and sucrose is one of the more soluble ones. At room temperature, a saturated solution, one that has as much sucrose dissolved as it can possibly hold, is two-thirds sucrose and one third water. And as temperature goes up, the percentage of the mixture that is sucrose at saturation also goes up. In hot tea, the water can hold up to three-quarters sugar, or more depending on the temperature.

We use this principle to make rock candy because sucrose can only crystallize from a supersaturated solution—that's a solution that contains more sucrose than the saturation concentration. The way we get to that supersaturated state is to play on the temperature effects of solubility. In fact, you could make rock candy from warm tea. When you dissolve the two packets of sugar into the warm tea, it's all dissolved. But when you cool it to iced tea temperatures, now there is more sucrose dissolved than what's allowed at saturation. It's supersaturated. That's probably what happened with that Chinese lady's syrup over time.

The next step in the process requires that the supersaturated sugar solution crystallizes. Somehow, the sucrose molecules in the liquid state, which are moving around randomly with a lot of energy, have to come together, settle down (losing their liquid energy), and organize into a crystal structure. That's where the string or stick comes into play. The rough surface provides an opportunity for the sucrose molecules to come together in clusters so they can organize into a crystal embryo, a process called nucleation. Nuclei can be thought of as the seeds from which crystals grow and in rock candy, we usually want those crystals to form on the stick or string, not on the bottom of the jar that we're using to make rock candy.

Once nuclei have formed on the stick, they begin to grow and continue to do so as long as the solution remains supersaturated.

The important thing here is that the crystals that formed on the stick grow as quickly as possible while minimizing formation of crystals on the bottom of the jar. Commercially, this involves a tricky balance between temperature and concentration to keep the supersaturation just right. At home it's not a big deal if there are crystals on the bottom of the jar.

Diabetics, don't despair. Up until now, no sugar-free versions of rock candy have been available. But apparently a company in Pakistan has figured out how to make rock candy from the main sugar-free candy ingredients—sugar alcohols (isomalt, xylitol and erythritol). In particular, isomalt, which is a hydrogenated version of sucrose, has similar crystallization behavior as sucrose, so it shouldn't be that hard to make isomalt rock candy.

Let's now compare the process for making rock candy with that for cotton candy (see previous chapter). For cotton candy, we take essentially pure sucrose, heat to melt, and then cool rapidly into the glassy state. The key is that cooling of the high concentration material is so fast that the sugar molecules don't have time to come together to form crystal nuclei. They just solidify in what random arrangement they had in the liquid/molten state. Crystallization, on the other hand, takes time. You can't rush rock candy, even the commercial products take several days to make.

When I hear Big Rock Candy Mountain, I envision a large mountain made of sugar crystals—huge mounds of rock candy piled on huge mounds of rock candy. But I'm not sure if it's really big "Rock Candy" Mountain or "Big Rock" Candy Mountain? Would a peppermint tree (or cigarette tree) grow on rock candy? Either way, a candy mountain, with lemonade springs and a soda water fountain, is just about candy heaven.

12

Candy Doctors

The word "doctor" has many definitions these days. Perhaps the first definition that comes to mind is the person who takes care of us, including the physician, who doctors our health, and the dentist, who looks after the health of our teeth. Perhaps it's fitting that both doctor and dentist may be pertinent in a book about candies, but the term doctor has a very different meaning to the confectioner.

As a verb, the word doctor can mean many things. It can mean to give medical assistance, to treat or apply remedies to, or to add a foreign substance to, as in to adulterate. Both physician and dentist fit under the first and second definitions; they provide medical assistance and treat our ailments. In confections, the term "doctor" has a different, very unique meaning, related to the third definition above.

A candy doctor is not someone who looks after the health of our candy, it's an ingredient added to a candy formulation that controls sucrose crystallization. In that sense, it fits the last definition in the list above, to add a foreign substance. The specific purpose is to control how sugar crystallizes, since that affects texture, appearance, taste, and a variety of other sensory attributes.

Why do we need to control sucrose crystals in candies? For one, sucrose crystals are what differentiate chewy caramel (no crystals) from fudge (highly crystallized). The easily broken structure of grained fudge is very different from the stick-to-the-teeth character of chewy caramel, a difference primarily due to the presence of crystals in one but not the other. Candy makers control this

R.W. Hartel and AK. Hartel, *Candy Bites*, DOI 10.1007/978-1-4614-9383-9_12,
© Springer Science+Business Media New York 2014

difference, at least in part, through addition of a doctor—the more doctor, the less crystals.

The first intentional doctor to be used in confections was most likely cream of tartar. It has an indirect effect—it doesn't actually provide any inhibition itself, but rather causes a true doctor to be created. In fact, cream of tartar is still added to some confections to provide control of sucrose crystallization. It works by causing the hydrolysis of sucrose, a disaccharide, into its component mono-saccharides, glucose and fructose. The acidity from adding cream of tartar, along with the elevated temperatures usually found in candy processing, leads to hydrolysis of the sucrose.

Hydrolysis of sucrose due to addition of cream of tartar creates a doctor, called invert sugar. The glucose and fructose molecules of invert sugar interfere with the ability of the sucrose molecules to come together and crystallize. To form crystals, sufficient numbers of sucrose molecules in close proximity and with sufficient mobility are needed. In crystallizing, the sucrose molecules organize with each other into a crystal lattice framework, each molecule taking essentially the same conformation.

Invert sugar, being made from molecules different from sucrose, gets in the way of this process. Not only are there fewer sucrose molecules (due to hydrolysis to form invert sugar) left to crystallize, but the glucose and fructose molecules impede the sucrose mole-cules from coming together to form the crystal lattice. More invert sugar, less crystals.

The presence of invert sugar moderates texture in partially crystalline candies like caramel, fudge, fondants and creams. To make chewy caramels, we use more doctors to prevent crystalliza-tion, whereas in fudge, where crystals are desired, less doctor and more sucrose is used.

Since controlling the exact amount of sucrose inversion with cream of tartar is difficult, it's no surprise that some enterprising company started producing invert sugar for sale. Confectioners could simply add a set amount of invert sugar, bought from the supplier, to their formulation to get exactly the proper amount of doctoring. Many old-time candy books still call for invert sugar as part of the formulation. However, invert sugar has largely been replaced in modern confectionery manufacture by corn syrup, the most common doctor in current use.

Corn syrup is made from corn starch. Starch molecules are long molecules made of a chain of glucose molecules (starch is a polymer made of glucose units). To make corn syrup, the starch molecules are cleaved to create smaller fragments. Complete hydrolysis of a starch molecule produces pure glucose. Partial hydrolysis results in fragments of glucose polymers with a wide range of size, from single glucose molecules to polymers with 15–20 glucose units. In fact, confectioners have a wide range of corn syrups available, with different molecular weight distribution.

Breakdown of starch can be accomplished with acid and heat, but can also be done using enzymes. In fact, the process is quite similar to what happens when you ingest starch. The enzymes in your mouth and saliva start to breakdown the starch, after which the acidic conditions in your stomach continue the process. Commercially, the two methods give slightly different types of corn syrups based on the mechanism for cleavage of the glucose polymer, but their effect on sugar crystallization is generally the same.

As a doctoring agent in confections, corn syrup provides enhanced protection against crystallization compared to invert sugar, primarily due to the presence of larger (high molecular weight) molecules. These partially hydrolyzed glucose polymers are better at inhibiting sucrose molecules from coming together as crystals than the smaller molecules found in invert sugar.

Corn syrup is often called glucose syrup (or sometimes just glucose) in other parts of the world. If the syrup comes from the starch in wheat or even potato, a more generic term is needed; hence, the term glucose syrup. In the US, glucose syrup comes primarily from corn starch so we use the term corn syrup almost exclusively. For those who want to show their global savvy, the term glucose syrup is coming more and more into vogue among confectioners.

But because Europeans simplify glucose syrup to glucose, it's not easy to differentiate between the syrup and the purified monosaccharide sugar, glucose. For this reason, the candy industry often calls molecular glucose by its other name, dextrose. It's pretty convoluted—almost need a scorecard, or at least a different type of doctor (a Ph.D.), to keep things straight.

It's funny how the candy maker needs so many doctors—the physician for health issues related to obesity (although the expression "never trust a skinny candy maker" is no longer so true these days) and the dentist to help fix tooth decay. But the most important one is the doctor that controls sugar crystallization.

13

LifeSavers or Jolly Ranchers

Urban legend has it that LifeSavers got their name because the daughter of the inventor died from choking on a piece of hard candy and thus, developed a candy with a hole in the middle. Truth or myth?

Before we get to the answer, let's compare Jolly Ranchers with LifeSavers—hard candies with different choking hazard potential. Besides the difference in shape, what other factors differentiate the two most popular hard candies?

Historically, LifeSavers are the older of the two. They got their start in 1912 as Crane's Peppermint Life Savers, a pressed mint candy that was intended to support the main chocolate business of inventor Clarence Crane. It wasn't until 1925 that the clear fruit drop LifeSavers candy with the hole in the middle was developed. The classic 5-flavor candy roll came out in 1935. Over the years, the LifeSavers brand has seen several owners, from Nabisco to Kraft to the current owners, Wrigley/Mars.

Jolly Rancher hard candies were first made by the Jolly Rancher Co. of Golden, Colorado. The company name was originally intended to suggest a friendly western image. They sold chocolates, soft-serve ice cream, and hard candy, although the hard candy side of the operation quickly became the highlight company product. Over the years, the Jolly Rancher brand also has been owned by various companies, and is now part of Hershey Foods.

Both candies were made in the United States, Jolly Rancher candies in Colorado and LifeSavers in Michigan, for many years. However, both are now made in Mexico or Canada to take advantage of world sugar prices (see Chap. 5).

R.W. Hartel and AK. Hartel, *Candy Bites*, DOI 10.1007/978-1-4614-9383-9_13, 49

Perhaps you prefer LifeSavers or Jolly Ranchers based on whether you suck on hard candy or crunch it. If the results of a recent informal internet survey have any validity, we're split on how we prefer to eat hard candy—there are about as many people who suck their hard candy as there are those who crunch it. But, that depends a little on which hard candy we're eating. LifeSavers, made with more sugar than corn syrup, are hard and crunch nicely in the mouth. Jolly Ranchers on the other hand are made with more corn syrup than sucrose, and don't crunch very well when you bite into them. In fact, even though Jolly Ranchers are classified as hard candy, they're often soft enough that the pieces stick to your teeth when you try to crunch them.

This difference in hardness and crunchiness between Jolly Ranchers and LifeSavers can be related to their differences in composition, which ultimately affect their glass transition temperature, where sugar glass begins to flow (see Chap. 9). Briefly, hard candies with higher glass transition temperature are more stable to moisture uptake and flavor loss, but are also more brittle.

We've measured the glass transition temperature of Jolly Ranchers to be slightly above normal room temperature (about 79 °F), whereas LifeSavers were found to have a higher glass transition temperature (about 108 °F). Since brittleness of a sugar glass increases with glass transition temperature, it's no wonder that LifeSavers crunch better than Jolly Ranchers.

However, the glass transition temperature also impacts flavor release, with lower values promoting more rapid release of flavors. That in part explains why Jolly Ranchers have such a powerful and immediate flavor hit—an important taste attribute in hard candy. In fact, LifeSavers have a small amount of high fructose corn syrup added to help speed up and enhance flavor release, but it's nowhere near that of Jolly Ranchers.

Another difference between Jolly Ranchers and LifeSavers is shelf life—what goes wrong with the candy over time and how fast that happens. What's the main problem with Jolly Ranchers? That sticky mess. You often have to peel the plastic wrap off the candy. Stickiness is due to the simple sugars found in Jolly Ranchers, which are also responsible for the low glass transition temperature. It doesn't take too long sitting in warm, humid air for Jolly Ranchers to get sticky, especially if the bag is open. The twist wrap packaging used for Jolly Ranchers is not a very good water barrier—water molecules can easily get around the twist in the wrap to make a sticky syrup layer on the surface of the candy.

In contrast, the 5-flavor variety of LifeSavers are much less prone to getting sticky because they have fewer small sugar molecules to pick up water, especially when they're packaged in the individual plastic wraps. However, because they're made with more sugar than corn syrup, they're prone to a different problem—graining, or sugar crystallization. When LifeSavers grain, the texture at the surface softens as crystals form. The outer surface becomes more like an after dinner butter mint than a sugar glass.

Another difference between the two candies is the organic acid used—LifeSavers use citric acid, Jolly Ranchers use malic acid. Citric acid is, not surprisingly, found in citrus fruit (lemon, lime, etc.) and complements citrus types of flavors. It has a sharp intense initial sourness that fades quickly over time. Malic acid, with a slower and smoother release, is found in apples and watermelon, so best complements these flavors. Neither candy is excessively sour—the acid simply acts as a flavor enhancer. Although you can find exceedingly sour hard candies (see Chap. 42), LifeSavers and Jolly Ranchers are on the very low side of candy sourness index.

So, is the urban legend about how LifeSavers got their name true? Are they really named LifeSavers because they were designed to prevent choking? No! The hole in the middle of the original LifeSaver candy, which was a pressed mint like Wint-O-Green (see Chap. 23), was necessary to help the tablet press operate more smoothly. In fact, the name, LifeSaver, originates from the similarity of the candy shape with the floating lifesavers being popularized after the sinking of the Titanic.

14

Candy Canes: The Science Experiment

What happens when you put a candy cane into the oven set at 350 °F? Go ahead and place several of them on aluminum foil on a cookie sheet in the oven, then come back and read this chapter.

The candy cane is said to have its origins at Christmas time in Germany circa 1670. A church choirmaster in Cologne gave sticks of hard candy with a crook at the end to the children in his choir to keep them quiet during long Christmas services. It's not clear why he chose the crook shape, perhaps it was indeed to signify the shepherd's staff, but he must have had a pretty good understanding of material properties to do it.

To make candy canes, a sugar syrup is boiled to about 300 °F, which drives water content down to just a few percent. On a candy thermometer, this would be the "hard crack" stage—if the hot sugar mass is dripped into a glass of cold water, it forms thin, brittle-hard strands of sugar candy glass (Chap. 8).

At high temperatures, the sugar mixture is sufficiently fluid that it can be poured onto a cold table. As the mixture cools, the candy maker adds mint flavor and periodically turns the mass until it reaches a semi-plastic molten state suitable for forming and shaping.

Making candy canes is a lot like blowing fine glass sculptures and intricate scientific glassware. In fact, pastry chefs make fine edible art from molten sugar in much the same way that artists make fine glass sculptures from heated window glass. When sufficiently heated, both window glass and candy cane sugar mass become sufficiently fluid so that they can be shaped and molded into any desired form. When cooled, both set into a solid, glassy

R.W. Hartel and AK. Hartel, *Candy Bites*, DOI 10.1007/978-1-4614-9383-9_14,
© Springer Science+Business Media New York 2014

state. The sugar molecules in candy canes have the same random molecular orientation as silica molecules in window glass.

How does the candy cane in your oven look now? After a few minutes at 350, it should still be holding its shape, although it should be getting soft. In fact, if you take it out at this point (careful, it's hot—you might want to use tongs and let it sit briefly before touching it), it's malleable and can be reshaped. You should even be able to twist the candy cane around itself to form a knot.

It's while in this physical state that the essentials of a candy cane are created during the manufacturing process, although that happens as the hot, newly-formed candy mass is cooling. To create the shiny white appearance of a candy cane, the semi-fluid, malleable candy mass is placed onto a pulling machine. Three arms rotate in a synchronous pattern to fold, stretch and refold the semi-plastic candy mass, just like in taffy pulling. Pulling incorporates tiny air bubbles, which provide the white, silky shine.

The original candy canes made by the Cologne choirmaster were all white. It wasn't until around the start of the twentieth century that a new twist appeared—in the form of red stripes. No one knows for sure why the stripes were added, but we do know how.

To add red stripes to the white stick of candy, a portion of the malleable candy mass is separated and red dye worked in. The red candy mass is further separated into several thin strips, which are "blocked" with thicker strips of white candy mass to form a stubby, log-like cylinder with alternating white and red stripes on the surface. The log is rolled on a table held at 170–180 °F to keep the candy mass in that semi-plastic, malleable state.

The ends of the molten log of red and white candy are pulled out to form a narrow rope, with candy-cane diameter, while still rolling it on the warm table. Barber-pole spirals of red color on a white background are formed with a slight twist of the rope, which is then cut into sticks of the desired length and bent over a wheel to make the crook. The key throughout the process is to keep the mass warm so it's pliable and easy to work.

With the red twists and crook complete, the candy cane is cooled quickly to room temperature to keep it from changing shape and to set the candy cane in the glassy state. If you tried bending it now, it would simply shatter and break, like any glass.

Candy canes were made by hand until 1950, when the brother-in-law of the owner of Bob's Candies in Atlanta, Georgia invented machines that could do it all: forming and twisting the rope, cutting the sticks, and even bending the crook. Modern candy factories twist out thousands of candy canes per hour using extruders—machines that shape the semi-solid mass through small openings—to put red striping on a rope of white candy. The rope is cut into sticks of candy cane lengths, after which each stick is given a rolling twist on a special machine. The final step is forming the crook, again over a wheel, but all done automatically.

Check the candy canes in your oven now. After about ten minutes, the sugar mass should become fluid enough to lose its shape and form a puddle of candy on the foil. This is the same principle used by glass-blowers, where the glass is heated above the point of melting, where it can be formed and shaped. If you now allowed the melted candy cane to recool to room temperature, it would once again become a glass, but no longer with the candy cane shape.

As a popular holiday treat, the candy cane's physical attributes make it an easy-to-hang, colorful addition to the traditional Christmas tree. It's also an intriguing way to teach the about the transition from a liquid to a glass, and back again.

15

Sponge Candy or Fairy Foam

Pumice is "a textural term for a volcanic rock that is a solidified frothy lava composed of highly microvesicular glass pyroclastic with very thin, translucent bubble walls of extrusive igneous rock." It takes a geologist to understand that description, most of us would call pumice an aerated rock.

Similarly, sponge candy may be called "a solidified frothy hard candy composed of highly dispersed bubbles supported by thin walls of amorphous sugar glass". It takes a candy scientist to understand that description, most of us would call it an aerated hard candy.

Pumice and sponge candy have much in common. For one, they look vaguely similar. Both have air pockets held by a solidified matrix (rocks in one and hard candy in the other) and are a creamy, brownish color. And they both float. In fact, a pumice "island" as large as Israel was recently seen floating off the coast of New Zealand. Speculation was that it came from an underwater volcanic eruption. Sponge candy floats too; they're both lighter than water due to the air pockets. Unlike pumice though, sponge candy quickly dissolves into a gooey mass after sitting in water for a minute or so.

Sponge candy has many different names. In Wisconsin, it's called fairy foam. On the West coast (and Michigan), it's known as sea foam. I don't know about you, but the term sea foam doesn't conjure up images of something I'd want to eat. It's more like that nasty frothy stuff we equate with some sort of pollution.

Around the world, names for sponge candy include honeycomb candy, puff toffee, and cinder toffee, a name reminiscent of pumice.

R.W. Hartel and AK. Hartel, *Candy Bites*, DOI 10.1007/978-1-4614-9383-9_15,
© Springer Science+Business Media New York 2014

My favorite name though is hokey pokey—that's what New Zealanders call it. I can only imagine hearing "hey mate, did you see that hokey pokey island floating off the coast?"

Although you can buy sponge candy at the store, it's relatively easy to make at home. All it takes is a pot and a stove to cook the sugar syrup. Equal parts of sugar and corn syrup, with some vinegar and a little excess water to dissolve the sugar crystals, are cooked to the hard crack stage (see Chap. 8). Sometimes brown sugar is used to add flavor. When cooked to a temperature of 300 °F, only 2–3 percent water remains in the syrup and, when cooled to room temperature, the sugar mass solidifies into hard candy, a sugar glass.

To make sponge candy, however, we need to aerate it just prior to cooling. That's where the baking soda comes in. As soon as it's done cooking, stir in the baking soda and watch it foam. The reaction between the acetic acid (vinegar) and sodium bicarbonate (baking soda) releases carbon dioxide gas, which foams the molten sugar syrup. If done correctly, the acid-base reaction occurs at the same time as the sugar mass is cooling, leaving a network of gas bubbles surrounded by walls of sugar glass, or hard candy.

Technically, there is a distinct difference between a sponge and a foam. Foam contains individual bubbles dispersed in some continuous phase (rocks for pumice and sugar glass for sea foam). The bubbles are not in direct contact; each one is separated by a layer of that continuous phase. In contrast, a sponge, in its generic use, not the undersea animal whose name we use, has lots of air cells (not bubbles) that are highly interconnected, but contained within a continuous phase. Some common foams include the head on a beer and meringue, while an example of a food sponge is bread, where the air cells are often interconnected. In this sense, it's probably more appropriate to call it fairy (or sea) foam rather than sponge candy since the individual bubbles are distinct, not interconnected.

Another somewhat similar candy is called divinity, primarily a southern treat. In sponge candy, aeration comes from the acid-base reaction whereas in divinity, aeration comes from whipping egg whites. To make divinity, the egg whites are whipped into a foam while the sugar syrup cooks, but only to the hard ball stage at about 265 °F. The hot syrup is then slowly folded into the egg whip and stirred until it thickens. The hot syrup causes the egg proteins to denature within the amorphous sugar mass as it cools. Once it sets, it can be cut into pieces and coated with chocolate.

When you first bite into sponge candy, it has a crunchy texture, but then the crunchy pieces quickly dissolve in your mouth, releasing the sweetness and cooked sugar flavor. That rapid dissolution in your mouth is reminiscent of cotton candy (see Chap. 10), another sugar glass with high surface area. Both candies are extremely hygroscopic.

That affinity for picking up water from the air makes sponge candy highly unstable. Water vapor molecules in the air quickly adsorb to the surface of the sugar glass, gradually penetrating into the thin layers of amorphous sugar that hold the bubbles in place. The increase in water content of the sugar matrix results in a drop in viscosity and eventually the mass flows and collapses. In fact, most recipes for divinity and sponge type candies say not to bother making it on a hot humid day. That must really limit when they can make it in the South.

To preserve fairy foam you have to keep it from the humidity. Probably the best way to do that is to coat it in chocolate. The layer of chocolate, a hydrophobic material since it's fat based, provides a decent water barrier and prevents the foam from collapsing except under the most humid conditions.

You may have been scratching your head thinking why does making candy sponge seem so familiar? Well, it's essentially the same reaction used in the classic science fair volcano to generate flowing lava. In the volcano, however, it's not a sugar-based candy but simply colored water made to look like molten magma when it comes foaming out of the volcano. As a twist to the standard science fair volcano scene, perhaps you can use fairy foam as a realistic and edible pumice substitute. Guaranteed to make you the most popular kid in science class.

16

Dum Dum Lollipops

What is the mystery flavor of a Dum Dum lollipop? The answer lies in processing technology and a creative spirit.

Dum Dum's are a popular example of the lollipop, or lollie, or sucker, or sticky pop, you name it. Hard candy on a stick. Instead of rolling the candy in your mouth as you suck on it, like a Jolly Rancher, the lollipop provides a convenient way to enjoy a sweet without the mess. Imagine taking the Jolly Rancher out of your mouth periodically to give your sense buds a rest. Whoever came up with the idea of putting candy on a stick helped make life less messy.

Lollipops have been around a long time, probably well back into the 1800s or earlier. However, it's thought that cavemen were the first to put a sweet on a stick—they used sticks to hold honey-based sweets. Perhaps they were worried about the mess too? Nah, probably not, they probably used the stick to dig honey out of the comb and then just ate it off the stick.

The name lollipop most likely originated in England. In one dialect, lolly means tongue, so a lolly pop is a lickable hard candy. Lollipops now come in a wide assortment of sizes, shapes and flavors. Some are flat, many are round, but many come in odd shapes, from the Chicken Sucker to the Banana Slug sucker. There are even erotic lollipops for adults. Some, like Dum Dums, are quite small, while others, like the all-day sucker, last all day. Many lollipops are filled with something in the center, from Tootsie Rolls to gum. Even scorpions and other bugs—that's gross.

The first modern lollipops were possibly made by accident, when a hard candy maker left a wooden stirrer in the pot. The

R.W. Hartel and AK. Hartel, *Candy Bites*, DOI 10.1007/978-1-4614-9383-9_16,
© Springer Science+Business Media New York 2014

remains of the sugar mass on the bottom of the pot solidified, leaving a blob of hard candy on the end of the stick.

The first intentional lollipops were made by first cooking the sugar syrup in a pot or kettle to remove water to reach the hard-crack stage (see Chap. 8). The molten, flavored mass was then poured into a mold. After the candy had cooled a little, it was sufficiently viscous that the stick could be inserted and remain straight, without falling against the side of the mold. When the candy cooled to room temperature, it became a solid candy, a sugar glass. The mold was opened and the candy removed. Artisan lolly-makers still use essentially the same process these days, by depositing the molten sugar syrup into shaped molds.

Large lolly-makers, like the maker of Dum Dum's, the Spangler Candy Company, use a large-scale continuous process. The Racine Confectioners Machinery Company (Racine, WI) claimed to produce the first automated lollipop production line in 1908, producing 40 lollipops per minute. Modern lollipop machines make many more than that.

While some candy companies use continuous sugar cookers that feed continuous lolly-formers, many candy makers, including the Spangler Candy Company, use older technology, a series of batch cookers that feed the lollipop line in a steady stream. Staggering the sequence of loading, cooking and unloading batch kettles allows a constant flow of molten hard candy mass into the process. Each kettle holds about 150 pounds of candy mass.

In fact, there are two steps involved in cooking one kettle of candy. In the pre-cooker, water, sugar and corn syrup are mixed together and cooked to a temperature around 270 °F. At this point, the sugar mass goes to a vacuum kettle to remove the rest of the moisture. Pulling a vacuum on the sugar mass allows water to be removed at lower temperatures; the reduced pressure allows water to boil at lower temperatures. Instead of cooking to over 300 °F to reach 2–3 percent water, the same water content can be reached while keeping temperatures below 280 °F. The lower temperature means that less browning occurs in the syrup, yielding a clear candy syrup prior to color addition.

The kettles feed the candy batch onto a cooling table, where the colors, flavors and acids are added. Adding flavors and acids during the cook stage would cause significant problems, like flavor loss and acid hydrolysis of sucrose. In Dum Dums, both citric and malic acids are used to provide tartness. The acids also enhance the fruit flavors. A series of plows push, fold and flatten the cooling candy mass into a plastic state, at which time it's fed to a batch roller. This device rolls the candy into what candy makers call a rope, a continuous snake of candy that is still sufficiently malleable that it can be formed. After going through a series of rope sizers to bring the rope to the proper diameter, individual candy pieces are stamped out in the characteristic ball shape of a Dum Dum. The pieces also get a stick inserted at this point, before being cooled to room temperature prior to packaging.

Over ten million Dum Dums come off this line every day, or about 2.4 billion each year. That's a lot of Dum Dums, enough to fill all the doctor's and dentist's bowls in the country, and beyond.

Dum Dums were first developed by the Akron Candy Company in Ohio in 1924. A sales manager dubbed it a Dum Dum as a name that any child could say. The Dum Dum brand was bought in 1953 by Spangler, who now markets 16 flavors plus the Mystery Flavor. Spangler periodically changes flavors to keep current, even holding a flavor challenge in 2012 to modernize the flavor portfolio.

What's that Mystery Flavor? Actually, it's changing all the time, but it's always a combination of 2 of the main 16 flavors under current production. It comes from the transition period when they're switching production from one flavor to another. There is a brief time during this switchover when Dum Dums contain both flavors. Rather than discarding or reworking this candy, as many candy companies do, Spangler found a creative alternative, to market the blended flavor as a mystery. Hence, the Mystery Flavor is nothing more than a blend of two flavors. It might be a Cherry/Cream Soda flavor or it might be Bubble Gum/Cotton Candy, whatever they happen to be making that day.

17

Cut Rock

Some hard candies have words or images in them. At the University of Wisconsin, we used to have mints made with the W insignia in the middle and the words UW-MADISON on the bottom. If you looked at it from the backside, the letters were reversed. It only reads correctly from one side.

Sometimes you can find disk-shaped hard candies with images of such things as roses, fruit slices or plants inside. Both the lettered and designed candies are called "cut rock", an incredibly artistic candy to make. Cut rock probably originated in England, where resort towns had a special "rock" candy that identified them. For example, the seaside resort of Brighton is known for its rock, sold most often as a cylinder of hard candy with the words Brighton Rock in the interior.

These are not the typical rock candy that we think of, bits of sugar crystals either in pieces or on a swizzle stick. This type of cut rock candy is really a hard candy that's been carefully constructed with the interior design in mind.

As a form of hard candy, the process for making cut rock starts as all hard candies start, by mixing the appropriate ratios of sugar, corn syrup and water. Once all the sugar has dissolved, the sugar mix is cooked rapidly to 300 °F, to the hard crack state (see Chap. 8). Since cut rock is typically an artisanal product, it's usually made in small batches. The batch of molten sugar mass is poured out onto a cold table where it's periodically turned in on itself to help promote cooling. The aim is to bring the sugar mass to a plastic state, where it's sufficiently malleable to work into shapes but sufficiently plastic that it will hold its shape for a little while.

R.W. Hartel and AK. Hartel, *Candy Bites*, DOI 10.1007/978-1-4614-9383-9_17,
© Springer Science+Business Media New York 2014

Since making cut rock takes time, perhaps 20–30 minutes for an experienced rock maker, the plastic candy must remain malleable longer than normal. This is accomplished by working it on a hot table, with warm water flowing underneath the metal surface. The warm temperature keeps it from setting up too quickly, extending the working time to create the interior designs.

Colors and flavors are typically added as the candy mass is cooling. Folding the candy over and over on itself helps disperse the colors and flavors uniformly through the mass. Cut rock is often mint flavored but any flavor can be used. Multiple colors are used to create the design. The lettering of cut rock is most often red surrounded by a white background, but any color combination can be used.

To create a white color, the plastic candy mass is aerated on a pulling machine. A hard candy puller, similar to a taffy puller, is a traditional device that pulls, stretches and folds the candy mass over and over again. It simulates the artisan candy maker pulling his candy on a hook, but instead is comprised of multiple counter-rotating arms. As one arm turns, it picks up the candy mass and folds it over on top of the candy being swung around on a second rotating arm. A stationary arm in the middle allows the candy mass to be pulled and turned over itself. Each fold incorporates air bubbles, which are then stretched as the candy is pulled. The result is a large number of small, stretched out air cells that scatter light and give the white color. Once pulled, that element of cut rock is returned to the warm table to retain its malleability until all the elements are put together.

While one person is pulling, a second candy maker has started "blocking" the lettering. To make letters, strips of red and white candy are arranged to create a block with the letter inside. An "I" is pretty simple, with only a few elements needed to make a blocked letter. Three strips of red-colored candy are arranged to form the letter, with strips of white-colored candy packing to make a block with the letter enclosed.

Other letters are more complex, especially letters like "A" and "P", which require irregularly shaped color blocks to be arranged carefully and "glued" together by lightly moistening the surfaces that come together. "A" for example requires a triangular white section for the inside and a rhombus-shaped white section for the bottom. These separate blocks of red that make up the lines.

Curved letters, like "C" or the top of the "P", are often made block-shaped to simplify the process. We call them blocks, but each letter is roughly a square of a couple inches on a side and about six feet long. Once made, the individual letter strips are put between blocks of wood on the hot table to keep them warm without deforming as the rest of the piece is constructed. To create a saying within the candy, the different letter blocks are arranged to spell out a word or phrase.

Creating the interior letters or design is really the artisanal part of making cut rock. Lettering is hard enough, but making an ornamental pattern, like a flower or watermelon slice, inside the candy is even more difficult and requires years of training and practice. To make a rose, for example, multiple colored strips are needed to create a colored flower and the green stem. These strips have to be placed in a design with patience and skill.

Once the interior designs are created, a cylinder of candy is constructed with a white core and the appropriately-stacked lettering or design blocks in the middle. A thin overwrap layer of colored, often red, candy is wrapped around the outside of the cylinder, which is now a foot or two in diameter and about six feet long. To make cut rock candy, this fat cylinder is carefully stretched out, reducing the shape of the interior design, often making it more legible. A rope with diameter of approximately an inch or so is pulled from the initial cylinder and sections of that rope cut to make the finished candy.

Arguably the most ornamental cut rock we've seen was made right here in our lab, by some artisanal candy makers who came in to teach a course. They made a candy with two crossed American flags in the interior. Without telling anyone what they were making, they separated and colored brown, red, white and blue sections of the candy mass. They then carefully blocked the strips of the different colors together in a preconceived design. When pulled out, the red, white and blue of two flags on crossing brown flag posts magically appeared.

Although cut rock is an old-time candy, it's making somewhat of a comeback. Besides being available at Christmas, when most cut rock is sold, it's now being used as favors at special events. Cut rock can be made as wedding favors, with the bride and groom's names separated by a heart. Probably the cutest cut rock candy is Baby Feet, designed as a birth favor, a much nicer gift than a cigar (although gum cigars are good too). Baby Feet contain a colored footprint (blue for boys and pink for girls); the little teeny-tiny toes are an especially nice touch.

18

Sugar-Free Candy

We had some rolls of sugar-free breath mints around the house that Scott, our young son/brother, got into one day. I don't know if he felt his breath needed even more freshening, but after he finished a first roll, he started in on a second. Within the hour he found himself racing to the bathroom, trying to relieve the pressure in his bowels. He learned a lesson, the hard way, about sugar-free candy that day.

Sugar-free candy has been around for many years, mostly as a replacement for diabetics, who can't tolerate sugar very well. Based on "alternative" sweeteners, primarily sugar alcohols, also called polyols, sugar-free candies have increased in popularity and consumption over the years for several reasons.

Defined as an alcohol with multiple hydroxyl groups (one oxygen and one hydrogen molecule, or OH in chemical terms), polyols are found in nature but are most commonly synthesized chemically. For example, sorbitol is found in some fruits at low levels, but is produced commercially by hydrogenation of glucose from starch. Most polyols are made by hydrogenating (adding a hydrogen molecule to) some type of sugar molecule. Other common polyols found in confections include maltitol (from maltose), xylitol (from xylose), mannitol (from fructose), isomalt (from sucrose), and, the tongue twister ingredient, hydrogenated starch hydrolysate (from corn syrup), sometimes called HSH for your tongue's sake. Maybe you've seen HSH listed on an ingredient deck but not known what it is—it's the sugar-free equivalent of corn syrup.

R.W. Hartel and AK. Hartel, *Candy Bites*, DOI 10.1007/978-1-4614-9383-9_18,
© Springer Science+Business Media New York 2014

Sugar-free candies have been gaining wider acceptance, but they generally suffer from a couple major problems that still limit their use. First, to many, they don't taste nearly as good as the original. Most people still consider sugar to be the best sweetener (apologies to our friends in the sugar-free business). Second, the sugar replacers typically are digested very slowly, if at all, meaning they pass right through and cause a laxative effect on their way. If you chew that entire pack of sugar-free gum, you'll find yourself cramping over and running to the toilet.

Why do polyols have a laxative effect? Because they're not absorbed in the stomach, they make it intact into our intestines where they draw water due to osmosis. It's actually a chemical effect related to the small size of most polyol molecules, known scientifically as the colligative effect. This same effect causes salted water to boil at a slightly higher temperature (boiling point elevation) and milk to freeze at a slightly lower temperature (freezing point depression) than water. The presence of the polyol in the gut causes an osmotic imbalance with the nearby cells, causing water to enter the intestinal tract in an attempt to balance the osmotic pressure. That extra water puts pressure on the bowels, eventually making us run to the bathroom to relieve ourselves.

In the days when the Atkin's Diet was the rage, when a carbohydrate was the nutritional devil itself, polyol makers were in heaven. Because polyols don't provide a glycemic or insulin response, which is why they're used in "diabetic" candy, they also fit nicely into the Atkin's program. Almost all foods were reformulated for the Atkin's Diet, from cereal to bread to candy, by replacing sugars and carbohydrates with polyols, with varying degrees of success. It was a boom time for the makers of sorbitol, maltitol, and the rest.

Polyols have some other interesting characteristics that either help or hurt them in certain applications. Being somewhat similar to sugars, polyols also have some sweetness, but just how sweet depends on the individual chemical make-up. Xylitol, for example, is just as sweet as sucrose, making it an excellent substitute in products like chewing and bubble gum. Maltitol is not quite as sweet, which is why sugar-free chocolates made with maltitol need to have some high-intensity sweetener added. Others may only be half as sweet, or less, as sucrose.

Since polyols are not absorbed very well by the body, they have lower caloric values. Sugars and carbohydrates in general provide 4 calories of energy per gram consumed, whereas polyols have somewhere around half of that, depending on the polyol. Polyols also help decrease the caloric count of foods reformulated to contain them. A sugar-free hard candy has about half the calories as its sugar-based counterpart.

Nearly every candy can be made sugar-free, but not all are commonly found at supermarkets. Sugar-free licorice, chocolate, creams, and jelly beans are all made, but the two candies that lend themselves most to sugar-free versions are hard candy and gum. In fact, sugar-free gum probably outsells sugar-based gum these days, and for good reason. One of the characteristics of polyols is that they don't support growth of the bacteria commonly found in your mouth and which are responsible for cavities. Besides, it's the only gum your mom will let you chew.

When you chew regular gum (or eat a pasta dinner or fruit for that matter), the sugars (broken down from starch for pasta) provide a food source for the oral bacteria to grow. This causes the pH to go down, which eats away the enamel. The result, a cavity, or carie as the dentists call it. Note that many foods are cariogenic, and the likelihood of a cavity increases the longer the sugars are in your mouth. Eat fruit or candy and brush your teeth immediately after, you'll get fewer cavities.

One of the real advantages of polyols is that they don't support growth of the oral bacteria, meaning they are noncariogenic (don't promote cavities). But one polyol, xylitol, is actually approved as being anti-cariogenic—it actually promotes dental health and the proof is so strong that FDA allows the claim to be written on the package. Besides not being used as a food source by the oral bacteria responsible for demineralization, xylitol has been found to actually promote remineralization of teeth, helping to fight against cavities.

One last trait of polyols that distinguishes them somewhat from sugars is the cooling effect when you eat them. When a sugar or polyol crystal dissolves, it takes heat out of the environment—that energy is used to increase the energy of the molecules as they transform to the liquid state. Most polyols have significantly more cooling effect than sucrose, with erythritol, a sugar alcohol produced by yeast fermentation of glucose, giving ten times the cooling effect as sucrose. That's great for a mint flavor, but not so good with most other flavors. That's unfortunate because erythritol has the lowest calorie content of any polyol, about 0.2 calories per gram (Europe allows that it has 0 calories).

While sugar-free candy has its place, it's not for everyone or all the time. Be careful eating too much of it or you too will be racing to the bathroom, trying to get there before the polyol in your gut does.

19

Pixy Styx and Fun Dip

Arguably the simplest "candy" is a handful of table sugar out of the bag. A really desperate kid, one who's got the sugar shakes, can simply stick a spoon in a bag of table sugar and get the blood sugar back to normal.

At Little League banquets when I was a kid, the adults sweetened their coffee with sugar cubes. I thought that was the best candy ever and raided everyone's table. Allowing a sugar cube to dissolve slowly in my mouth was a lot more satisfying to me than pouring a packet of loose sugar crystals down my throat. Yet, there are several commercial candies that are just that—powders that you either pour down your throat or dip with some wetted utensil.

It doesn't take much to make a candy powder out of table sugar. Sugar crystals, with a bit of color, flavor and maybe some acid to give tartness, are all it takes to make an interesting product that's popular with kids. Colors and flavors can be sprayed onto the sugar crystals followed by a drying step to create the powdered candy.

Actually, many of the powdered sugar candy products are made with a sugar other than sucrose (commonly called table sugar). Look at the label of candies like Pixy Stix, Fun Dip (also known as Lik-m-aid) and other dipping candy powder products and you'll see that dextrose is typically the main ingredient. Dextrose is the industrial term for glucose, a monosaccharide that's found naturally in fruits. Although there is a scientific basis for calling glucose dextrose, supposedly the term dextrose was popularized because the term glucose conjured up images of glue. Who wants glue-cose in their candy?

R.W. Hartel and AK. Hartel, *Candy Bites*, DOI 10.1007/978-1-4614-9383-9_19, © Springer Science+Business Media New York 2014

The glucose used in the candy industry comes from corn, not glue, but we'll leave that story to another chapter (see Chap. 12). Companies use dextrose in powdered candies because it flows nicely and doesn't cake very readily. It also has sort of a cooling effect in the mouth as it dissolves, a property that enhances the eating characteristics of powdered candies.

Pixy Stix come in a tube—open one end and pour it into your mouth. To eat Lik-m-aid, or Fun Dip, another classic powdered candy (made in Loompaland by Willy Wonka), you can use either your fingers, like we did years ago, or use a sugar stick supplied with the candy, the more common way now, to dip out the powder. Once your finger or the stick is wet, the powder adheres nicely and you can lick the sweet stuff off. I suppose they added the dipping stick because parents complained about how messy it was to eat Lik-m-Aid with your fingers, not to mention that it might not be very sanitary. Another such product are Baby Bottle Pops, where it's a lollipop in the form of a kids nipple (or nookie?) that gets dipped into the powder, so it looks like you're sucking on a bottle.

Pixy Stix also come in huge tubes for those who can never get enough. At 21 inches high and 4 ounces of sweet and sour, what more do you need for a Pixy Stix eating contest? Once, a couple of AnnaKate's friends, under the influence of girlish peer pressure and Coke, decided to see who could eat 25 of the little sticks the fastest. They poured the contents of the sticks into their mouths, but their giggles caused the powder to go down their windpipes and they raced for the trash can. Their coughing fits were so violent that the whole sugar mix came right back up again.

Despite being a pretty simple candy, there's still some interesting science related to candy powders. For example, there's the concept of deliquescence, which relates to how crystals interact with water. In fact, the relationship between water and sugar crystals is quite complex, although we'll only scratch the surface, so to speak.

Deliquescence is one cause for a bag of free-flowing sugar powder to turn into a hard brick that doesn't even break apart with a sledgehammer. It's bad enough if it's a one pound bag of sugar, but it's really bad when it happens to a tote containing a ton of powder. Due to deliquescence, a powder that should simply flow out by gravity has cemented together and can only be cleaned out with a jackhammer. That's not a good thing, perhaps bad enough for someone to lose a job.

Technically, deliquescence is defined as "tending to undergo gradual dissolution and liquefaction by the attraction and absorption of moisture from the air." When sugar crystals are exposed to air, water molecules in the air (water vapor) adsorb to the surface of the crystal. With more water molecules in the air (higher relative humidity), more water vapor adsorbs to the surface. The deliquescence point is reached when there are sufficient water molecules on the surface to actually dissolve some of the crystal to form a layer of sugar syrup.

When the humidity goes back down, usually with cycling temperature, the syrup layer dries out again as water molecules now want to go back into the air. If two crystals that have formed a syrup layer at high humidity are in contact within the bag of sugar, the dried syrup layer can cause the two crystals to fuse. With enough humidity cycles, what were once all individual crystals in a tote of sugar become one single fused agglomerate.

In Pixy Stix, both glucose crystals and citric acid crystals each undergo deliquescence individually. But what's really interesting, to a science geek anyway, is that the mixture of the two crystals actually enhances deliquescence of each other, making the mixed powder even more prone to form clumps. The science behind this is still the subject of study, but it relates to how the water molecules share between the two different crystals in close proximity. The pharmaceutical industry is keenly interested in deliquescence as well, since keeping powders free-flowing is critical to proper drug dosage.

If done right though, the packaging is sufficient to protect the candy from the worst of summer humidity and the shelf life of a Pixy Stix is over a year. Thanks to packaging engineers, deliquescence doesn't turn Pixy Stix powder into a solid Pixy stick.

20

Pez

PEZ candies are a nostalgic treat for adults and a super fun toy for kids. Perhaps the original "play with your food" idea, PEZ combines two of kids favorites, candy and toys, into one interactive dispenser. Part of the fun of eating PEZ is picking out the dispenser. In fact, some people buy PEZ primarily for the dispenser. What's your favorite PEZ dispenser?

PEZ was developed in the 1920s in Austria as a breath mint. In fact, the name derives from the German word for peppermint, pfefferminz, using the first, middle and last letters to spell PEZ.

Edouard Haas III, inventor of PEZ, was somewhat of a health freak who thought sucking on a breath mint was much healthier than smoking. One of the earlier advertising slogans for PEZ was roughly translated as "Smoking prohibited. PEZing allowed." Using a well-known advertising approach to build his business, he hired attractive PEZ girls to promote the breath mints throughout Europe.

PEZ were sold in rectangular tins until 1949, when Haas developed a box with a hinged lid, meant to mimic a cigarette lighter, to dispense the candies. Besides being a trick on smokers who asked for a light—here, have a mint instead—the dispenser was a hygienic way to share the mints.

PEZ, the breath mint, was brought to the United States in 1953. When the breath mints didn't catch on, PEZ-Haas developed a fruit-flavored candy tablet to market to children. At the same time, toy-like dispensers were developed to further attract kids. The combination of toy and candy became a huge hit, one that

R.W. Hartel and AK. Hartel, *Candy Bites*, DOI 10.1007/978-1-4614-9383-9_20,
© Springer Science+Business Media New York 2014

still enjoys a huge following these days. More than three billion PEZ candies are sold in the United States each year.

An example of a pressed tablet, PEZ candies are made by compressing sugar crystals under high pressure so the particles bind together. Other examples of pressed tablet candies include Wint-O-Green LifeSavers (see Chap. 23), Smarties, SweeTarts, and Runts, among numerous others.

Tablet candies are made primarily from a flavored sweetener base that is granulated to promote cohesion under pressure. In PEZ, sucrose crystals make up the base. Unfortunately, sucrose crystals don't compress very well by themselves and some preliminary steps are needed to get sucrose to form a coherent tablet. The process is called wet granulation. A liquid binder, in this case corn syrup, is mixed with the sugar crystals to make a paste, which is then pressed through a screen to make shreds of Play-Doh-like candy (sort of like making a Play-Doh hamburger). The shreds are dried, ground into a fine powder, and sieved to the proper size. This granulation is comprised of particles that themselves are made of numerous small sugar crystals held together by the binder, which promotes cohesion during compression. When compressed in a tablet press, these sugar aggregates fuse together very nicely to form a hard tablet.

Some tablets are made with other sweeteners. SweeTarts and Runts are made with dextrose while the sugar-free mints like BreathSavers are made with sorbitol. Because of the nature of dextrose and sorbitol crystals, neither one has to be wet granulated to form tablets under compression. The main concern with these candies is that the powder formulation has the proper particle size distribution to compress effectively.

The process of compression is used for other products besides candy tablets. Things like charcoal briquettes, animal feed pellets, and especially pharmaceutical tablets are made in the same way. In fact, tablet presses, even those used for candy, are highly regulated because of the pharmaceutical applications. In the wrong hands, they can be used to make illicit drugs.

A tablet press has several functions, all designed to rapidly and efficiently produce a finished tablet product. A hopper controls the flow of the dry powder granulation, which is fed into a cylindrical chamber between two punches, top and bottom. A feed bar sweeps the top of the chamber clear milliseconds before the two punches come together, applying pressure to create the tablet. Once formed, the two punches come apart and the bottom punch lifts to eject the tablet from the press. A rotary press has a series of punches operating sequentially, essentially spitting out tablets at a rapid rate. At hundreds of tablets per minute, a tablet press is an efficient, but noisy, piece of equipment.

Under pressure, the particles in the granulation deform and flow to fill in the spaces between them. A free flowing powder has a bulk density much less than the density of any individual particle because of the air spaces that separate them. Spheres of the same size can only be packed to a maximum limit, when carefully arranged in a tetrahedral arrangement, where about 74 percent of the volume is solid and the rest is air space. A packing density above 74 percent isn't possible, and usually bulk density of a powder is a lot less than that. When the powder is randomly filled into the press die, the bulk density is probably only about 60 percent, with numerous air spaces between the particles. After compression, the packing density approaches 95 percent or higher, as particles are forced by the pressure to fill in nearly all the spaces. It's this packing that provides the hardness of the tablet. The fewer air spaces, when using higher pressure, the harder the tablet.

To help control press operation as efficiently as possible, candy tablet makers use a lubricating ingredient to keep the freshly formed tablet from sticking to the wall of the press. Look at the ingredient list of any compressed tablet candy, like PEZ, and you'll often see either magnesium or calcium stearate. This wax-like material coats the particles of the granulation and allows the tablet to escape from the tablet press without incident. This is especially important at the high speeds attained by modern tablet presses. When used at too high a level though, these compounds actually work against formation of the tablet, so they're kept well below 1 percent.

There is some discussion about calcium and magnesium stearate having potential negative side effects, but these concerns are unfounded. Both are simply minerals combined with a fatty acid, both already prevalent compounds, even necessary compounds, in our diets, and beyond. Both calcium and magnesium stearate are forms of soap, nothing more than the salt of a fatty acid. Although typically not found in commercial soaps, calcium and magnesium stearate are found in the soap scum that forms when regular soap meets hard water (the source of magnesium and calcium). Although soap scum is kind of gross, there are no problems with these stearates at the levels used in tablets.

But interestingly, PEZ does not use magnesium or calcium stearate. The ingredient list for PEZ includes vegetable fats and an emulsifier as the lubricants. Another tablet confection, of sorts, Fizzies (see Chap. 21), also uses a different lubricant; in Fizzies, wheat germ oil is used to help ensure lubrication between the tablet and the die press.

Are you a collector of PEZ dispensers? According to an often repeated myth, the founder of eBay created the online auction site so his fiancé could more easily trade her favorites. However, this story was created by a public relations manager in 1997, two years after the site was created. While the tale was erroneous, the fact that people thought it was plausible shows how coveted these plastic dispensers can be.

What's the weirdest PEZ dispenser you've seen? How about the giant Death Star dispenser of giant PEZ? Or the largest PEZ dispenser, a seven foot ten inch snow man (not officially blessed by PEZ though) at the PEZ Museum in California?

21

Fizzies

Will the third time be the charm? Will the current version of Fizzies succeed where the previous two didn't? Only time will tell.

For those who recall, the original Fizzies came out in the mid 1950s, an invention of the Emerson Drug Company, who also created Bromo-Seltzer. A tablet candy of sorts, the intent was to create a bubbly drink by simply plopping the tablet into some water. Plop plop, fizz fizz, make your own soda pop.

The original Fizzies were a victim of the FDA. They were sweetened with cyclamate, a high intensity sweetener (30–50 times as sweet as sugar) that was suspected, falsely as it turns out, of being a carcinogen. When the FDA banned cyclamate in 1969, Fizzies was out—the demise of the first version.

Sodium N-cyclohexylsulfamate, or cyclamate, is the chemical name for the original sweetener. It was found by accident, by a grad student who was smoking a cigarette in the lab while working on the synthesis of an anti-fever medication. He put his cigarette down on the edge of the lab bench, apparently close enough to the chemicals he was working on. When he took his next puff, he discovered something sweet. Things were a lot looser and less controlled in chemistry labs back then. Suppose that chemical had tasted bitter or, worse yet, had been poisonous? One deadly example of old-time chemistry is Carl Scheele, a well-known chemist who died in 1786 from sniffing and tasting all the chemicals he worked on.

Cylcamate was banned as the result of a study done in 1969 that suggested that cyclamate, or actually a cyclamate-saccharin mixture that was being used as an alternative sweetener, caused increased

R.W. Hartel and AK. Hartel, *Candy Bites*, DOI 10.1007/978-1-4614-9383-9_21,
© Springer Science+Business Media New York 2014

incidence of bladder cancer in rats. Turns out the level tested would require drinking 350 cans of diet soda sweetened with that mixture. Still, in the interest of protecting our health, that was enough for the FDA to put the ban on cyclamate. However, it turns out that study could not be reproduced and 55 countries, including Canada, currently approve its use as a sweetener. But not the United States. The FDA now has ruled that cyclamate is no longer implicated as a carcinogen, but still has not lifted the ban.

Why did Fizzies use cyclamate in the first place, why not just use sugar? Turns out that in order to get the equivalent sweetness from sugar requires a Fizzies tablet as big as your head. Not a real practical solution; hence, the high intensity sweetener. Back then, there weren't as many choices for alternative sweeteners and cyclamate was a good choice, at least for a while.

The second try for Fizzies was in 1995 when they were reformulated using aspartame. Aspartame is another high intensity sweetener, with about 200 times the sweetness of sugar. The owner of the trademark approached Amerilab Technologies in Minnesota, specifically to produce the new Fizzies and bring them to the market again. Unfortunately, the second coming of Fizzies also fizzled out after a year or two, when the trademark owner's company went out of business. No FDA this time, just not a successful business endeavor.

The third coming of Fizzies was in the 2000s. The CEO of Amerilab Technologies bought the trademark since he couldn't imagine a Fizzie-less world. In this version, the sweetness is provided by the combination of acesulfame potassium (Ace-K) and sucralose.

Ace-K, at 200 times sweeter than sugar, was another serendipitous discovery, through another chemistry lab mistake. In 1967, a chemist accidentally dipped his fingers into the chemicals he was working with and then licked his fingers to pick up a piece of paper only to discover that his chemical was sweet. Ace-K is also suspected of having safety issues, but is approved for human consumption by the FDA. Sucralose, a chlorinated sucrose molecule 600 times sweeter than sugar, was discovered in yet another chemistry lab mishap. The chemist was asked to test the compound, but he misunderstood and thought he was being asked to taste it. So he tasted it, and it was sweet. Sucralose also has its detractors who think there are safety issues, although as with Ace-K, the FDA approves it for food use.

Besides the artificial sweeteners, Fizzies contain an acid, citric acid, and several basic salts, potassium and calcium carbonates, and potassium and calcium bicarbonates. When Fizzies are plopped into a glass of water, the acid-base reaction is initiated by the water and carbon dioxide bubbles are produced. This reaction is the same as that used for making candy sponge or fairy foam (see Chap. 15) and the foamy magma in science fair volcanoes. In this case, the acid-base reaction provides the fizzy goodness in your drink. When you drop the tablet into the water, watch the rising streams of carbon dioxide bubbles cause the tablet to bounce around as it dissolves. Cool.

Alka Seltzer, first introduced in 1931 as a cure-all for anything that ailed you, was probably one of the inspirations for Fizzies (along with Bromo-Seltzer). The fizzy reaction is the same one that occurs with Alka Seltzer, which also contains sodium bicarbonate and citric acid. While Alka Seltzer contains aspirin for a headache, Fizzies contain a sweetener for enjoyment. Bromo-Seltzer is similar, containing acetaminophen, sodium bicarbonate and citric acid.

Fizzies, Alka Seltzer and Bromo-Seltzer are all made in the same way, by pressing the powders under pressure until the powder particles fuse into a solid tablet, in the same process as used for Pez and SweeTarts (see Chap. 20). Crystalline powders of citric acid and sorbitol are blended with the salts, colors and flavors, a lubricant, and an anticaking agent. The mixed Fizzies powder flows into a die where it's squeezed under high pressure by both an upper and lower punch. In a continuous motion, the powder enters the die, the punches come together to form the tablet, and then the tablet is released as the punches separate. The current version of Fizzies also contains small amounts of wheat germ oil and magnesium oxide. The magnesium oxide, an anticaking agent, helps the powder flow into the die prior to compression while the wheat germ oil helps the newly formed tablet release from the die as the punch is removed.

Will this version of Fizzies be more successful than the previous versions? Will it last more than a few years? Both Alka Seltzer and Bromo-Seltzer have been around for a long time and there's no reason to think this version of Fizzies will not be just as successful.

22

NECCO Wafers and Conversation Hearts*

In grade school, one of the best days of the year is Valentine's Day. Everyone makes colorful construction paper mailboxes, which always fill to the brim with notes of affection. Treats are brought, and consumed, in large amounts, without any parents around to say, "No." Even a simple conversation heart is enough to make even the most cynical of kid squeal with delight. But children grow up and many find conversation hearts are no longer a source of delight. It takes a special kind of person to love these candies into adulthood.

The conversation heart is the seasonal cousin of the NECCO wafer, the bane of every child's Halloween bags. What makes these two candies so often hated stems from their texture, which some people liken to eating chalk.

The process of making both the NECCO wafer and the conversation heart is so simple, you can do it in your kitchen. First, corn syrup, sugar, gelatin, and gums are mixed with water to make a binder solution. Colors and flavors can also be added.

The liquid binder, so called because it holds the sugar crystals together in the candy, is slowly added to finely powdered sugar until the mass reaches a consistency like Play-Doh. It is then rolled out into a sheet and a die is used to cut out any shape desired, similar to making Christmas cookies.

A holey webbing of dough remains after the shapes are stamped out. This can be collected and reworked into the next batch. On carefully designed conveyors, the webbing is lifted away and sent back to the dough mixer to be reused. The candy that remains on

---

*Primarily by AnnaKate Hartel

R.W. Hartel and AK. Hartel, *Candy Bites*, DOI 10.1007/978-1-4614-9383-9_22,
© Springer Science+Business Media New York 2014

the conveyor is passed into a dryer to remove most of the water, leaving a hard durable sugar candy piece. A similar process is used to make Altoids.

Candy buttons and the more controversial candy cigarettes can also be made with a similar dough. Instead of punching out shapes, simply drop a small amount onto some paper or roll it into cylinders. It may be hard to find these old timey candies in stores now but both are available to purchase online. Luckily, this limits a child's exposure to the possibly harmful candy cigarette (do you really think that eating candy cigarettes as a kid leads to a smoking habit?)—only someone with a credit card can buy them.

The innovation that brought the NECCO wafer into widespread distribution was the cutting machine, invented in 1847 by Oliver Chase. Using a simple hand crank, the dough is cut into uniform discs in a matter of seconds. Back then they were called hub wafers. The brand name NECCO didn't come into usage until 1912, after a merger of smaller companies produced the New England Confectionary Company (NECCO).

Twenty years after the hub wafer was introduced, Oliver's brother Daniel used a heart-shaped die and edible ink to create the conversation heart. Then, the candy was given away at weddings because they were printed with popular proverbs about weddings and marriage. Since then, the sayings have been updated annually to reflect the changing culture of our society. Although some of the old standards, like Kiss Me and True Love, are consistent each year, modern sayings, like Tweet Me and Text Me, are continually being added, sometimes through consumer contests. Other sayings, like Dig Me and Dream Team, are dropped as they become outdated.

The original eight flavors of NECCO wafers—lemon, lime, clove, cinnamon, watermelon, licorice, and chocolate—can, today, cause the scrunched face of the thoroughly disgusted. For anyone that's had an impacted wisdom tooth, anything clove flavored has particularly negative associations.

Recently, NECCO chose to replace the original flavors and colors with natural versions. Because of the limitations of truly certified natural products for coloring and flavoring, the flavors had to change slightly. The chocolate flavor became a stronger, more intense cocoa flavor, the cinnamon flavor was less like Red Hots, and the lime flavor had to be dropped completely, leaving only seven flavors in the roll. Within a couple years of the change, sales had plummeted. Like many companies that try to reformulate an old product, NECCO had not anticipated the backlash from those who loved the original. After they received enough consumer complaints, they brought back the originals and sales returned to normal.

The types of flavors that can be used in NECCO wafers and conversation hearts are very limited. The packed sugar crystals and intense sweetness overpowers most flavors. Do NECCO wafers really taste like lime or watermelon? Not really. Those delicate flavors don't come through the chalky texture very well. Even stronger flavors like chocolate and licorice still taste more like chalk than anything else.

In fact, the flavor that works best in these candies is mint. Altoids are a good example of a similar product, but one that actually tastes good. The mint flavor is strong enough, curiously or not, to offset the chalky texture of these candies. Still, enough people enjoy conversation hearts and NECCO wafers that tons of them continue to be sold each year.

So what gives these candies their distinctive crunch? The amount of water used to make the dough has a large impact on the texture. Adding more water dissolves more sugar crystals, which produces softer dough. Then during drying, the sugar recrystallizes to form bridges, which makes a strong network of tiny sugar crystals. That's what gives it the crunchy, chalky quality.

But the texture of these little wafers comes with an upside. The hardness of this type of candy means it can last a long time—over five years. They don't lose flavor, they don't pick up moisture or dry out, and they don't support mold growth. There's really not much that can go bad about them. This robustness is why they were part of soldier's rations from the Civil War all the way to World War 2. These little wafers are so hardy that Admiral Richard Byrd took

23

Wint-O-Green Mints

Turn off all the lights and, while facing a mirror to watch yourself, bite into a Wint-O-Green Lifesaver. Or find a partner, look into each other's mouths in the dark and chomp a Wint-O-Green. You'll see flashing lights. No, it's not a hallucination, or magic, it's candy science.

The phenomenon is called triboluminescence, caused by electrons being released into the air as sugar crystals are broken apart by your teeth. In fact, many things release triboluminescent light when either broken or pulled apart. Even ripping duct tape off a counter generates a spark, although too small to see easily. Other candies, from NECCO wafers to Altoids, give the effect as well. Even other LifeSavers spark a little when cracked, but not as clearly and distinctly as the Wint-O-Green flavor. Try the above experiment with a NECCO wafer or a Butter Rum LifeSaver—you'll be disappointed.

LifeSavers are over 100 years old, with the first version, Pep-O-Mint, being developed in 1912. Clarence Crane, the inventor, a chocolate maker, was looking for a candy that would hold up better in the summer heat and decided to press sugar together into a tablet. The Wint-O-Green flavor wasn't brought out until 1918, but they've been popular ever since as a breath-freshening mint that withstands almost all storage conditions. If you stashed them away for months in a pocket or drawer somewhere, they'd still be good to eat. Pretty much indestructible. Inventor Crane was successful in his attempt to make a stable candy.

LifeSaver is a good example of a candy brand that has been bought and sold numerous times over the past century. Wrigley is

R.W. Hartel and AK. Hartel, *Candy Bites*, DOI 10.1007/978-1-4614-9383-9_23,
© Springer Science+Business Media New York 2014

the current owner, purchasing the brand in 2004, although technically it's owned by Mars since they took over Wrigley a few years
ago. The new management of the brand has led to numerous new
product introductions around the LifeSaver brand. Besides developing a host of new flavors, including the sugar free Fruit Tarts,
Wrigley has also expanded on the LifeSaver gummy category to
play on the long-standing success of the 100-year old candy.

The ingredient list for Wint-O-Green LifeSavers is very simple. They contain sugar, corn syrup, artificial flavor (primarily, oil of
wintergreen, although it doesn't specifically state that on the label),
and stearic acid.

Although considered a hard candy, Wint-O-Green LifeSavers
are not a boiled sweet like most hard candies. Regular LifeSavers
are cooked to the hard crack stage and cooled into a glass to make
hard candy (see Chap. 8). Wint-O-Green LifeSavers, on the other
hand, are a compressed tablet, where high pressure converts a sugar
powder into a discreet tablet (see Chap. 20). Both candies are
definitely hard, but Wint-O-Green LifeSavers are hard because
of the bonding between powder particles when forced under pressure. How can you tell they're compressed tablets? The stearic acid
in the ingredient list is a dead giveaway. Although many tablet
candies use calcium or magnesium stearate as the lubricant to help
the tablet eject cleanly from the press, Wint-O-Greens use stearic
acid as the lubricant.

Stearic, or octadecanoic, acid is a fatty acid consisting of eigh-
teen carbons in a straight chain. It's most commonly found in
natural fats including cocoa butter, tallow and lard. With a melting
point of 157 °F, it fits the requirements of a lubricant in a tablet
candy. It coats the particles as they're squeezed together and that
allows the tablet to slide smoothly across the metal surfaces of the
press without binding or sticking.

When you bite into a Wint-O-Green LifeSaver tablet, the
sugar crystals themselves are broken apart. This releases free electrons, which then impart their energy to the nitrogen gas molecules
in the air. The excited nitrogen molecules release the extra energy
primarily as ultraviolet radiation, although some light is emitted in

the blue region of the spectrum as well. Lightning during a thunderstorm is essentially the same mechanism, electrons exciting nitrogen molecules in the air. In that sense, when you crack a Wint-O-Green LifeSaver with your teeth, you're creating a tiny lightning storm in your mouth.

The phenomenon of triboluminescence has been known for centuries. In 1605, two years before he was knighted, Sir Francis Bacon noted that cane sugar released light when crushed. Besides being the first to document triboluminescence, Sir Francis Bacon is credited with numerous scientific advances. Specifically, he is credited for developing what we know today as the scientific method. For this, he is sometimes referred to as "the father of experimental science."

Scientists believe that in order for a crystal to release an electron when crushed, the molecular arrangement of the molecules in the crystal lattice has to be asymmetric, or at least contain sufficient impurities to cause asymmetry. Other crystals that exhibit triboluminescence are diamond and salt. In fact, when diamonds are cracked, or even rubbed vigorously, they produce a red or blue color as the electrons are released to react with nitrogen in the air. Well, that's what other people say—I haven't gotten close enough to a diamond to do the experiment myself.

We've noted that the light emitted from crushing sugar crystals occurs primarily in the ultraviolet (UV) region of the spectrum. Humans don't see in UV light so we miss the major flash of the Wint-O-Green LifeSaver. In fact, UV light is bad for our eyes. Staring directly into the sun allows UV light access to the retina, causing severe damage and even blindness. Other animals, however, can "see" in UV light. Birds, bees, butterflies, and especially the mantis shrimp, all use the shorter wavelengths of UV light to illuminate their way. Only one mammal has the ability to see in UV light—the reindeer. Maybe snow blindness led to their adaptation to UV light, but it allows them to see the lichens they eat in winter as well as the urine trails of their predators.

If we don't see in UV light, how can we see triboluminescent sparks in Wint-O-Green LifeSavers? The answer is in the flavoring. Oil of wintergreen contains methyl salicylate, which is responsible for the blue sparks visible when a Wint-O-Green LifeSaver is cracked. Methyl salicylate is also used in a variety of other products, including topical analgesics (Bengay) and mouthwash (Listerine).

Methyl salicylate also absorbs the UV light created from the electrons during triboluminescence, and then re-emits that light in the blue region of the spectrum, where we humans can see it. So, the reason Wint-O-Green LifeSavers are world renowned for their sparkle in the dark is the flavoring.

24

Peppermint Patties

Was Peppermint Patty, the Peanuts character who is an outstanding baseball player but only a D-minus student, named after the chocolate-covered mint patty originally made in York? With tomboy traits and freckles, she first appeared in the comic strip in 1966. She is perhaps best known for coaching the baseball team that always beat Charlie (she calls him Chuck) Brown's team. That, and not always having a good grasp on reality—"that funny kid with the big nose is a beagle?"

It's possible that the Peanuts character was named after the candy since York Peppermint Patties originated in York, Pennsylvania, in 1940 at the York Cone Company. Originally, the company produced ice cream cones and waffles, but when owner Henry Kessler set out to develop a new candy that combined chocolate and peppermint, the Peppermint Pattie was born. The early success of the Peppermint Pattie led the York Cone Company to drop their other product lines to focus on the chocolate-covered peppermint cream.

One of the defining characteristics of the original chocolate-covered mint cream was its snap. The original cream patty was hard enough that it would snap. In fact, legend from York has it that when Patties didn't meet the snap test, they wound up in the "seconds" pile and were made available to the townsfolk.

Another notable peppermint patty is made by Pearson's Candy Company from Minneapolis, Minnesota. Originally formed in 1909 by the Pearson brothers as a distribution company, they put out their own candy, the Nut Goodie Bar, in 1912. The Pearson's Mint Pattie first appeared in 1951, through acquisition of the

R.W. Hartel and AK. Hartel, *Candy Bites*, DOI 10.1007/978-1-4614-9383-9_24,
© Springer Science+Business Media New York 2014

Trudeau Candy Company. Although Pearson's still makes other candies (regional favorites like the Nut Goodie, Bun Bars, and Nut Rolls, as well as the newly acquired Bit-O-Honey, see Chap. 54), the Mint Pattie is the main reason people know the brand. Pearson's also acquired other candy brands, for example when they bought the Sperry Candy Company in 1962. With this acquisition came such candy favorites as the Chicken Dinner Bar and the Denver Sandwich, both on the dead candy list (see Chap. 64).

A Peppermint Pattie is technically a fondant cream, made by whipping together sugars with fat and flavor, forming a patty, and then coating it with chocolate. There are numerous ways to skin the cat, or coat the patty, so to speak. To make peppermint patties at home, you could cream butter with powdered sugar and peppermint flavor (a little corn syrup and/or cream are also sometimes included) until light and creamy. Alternatively, some recipes call for adding powdered sugar to sweetened condensed milk to make a stiff dough rather than creaming the fat and sugar. These patties will be a little firmer. After allowing the dough to set briefly in a freezer or refrigerator, the patties are cut from a sheet of the cream, just like making Christmas cookies. The patties are then hand-dipped in tempered chocolate and laid out for the chocolate to solidify.

As is often the case, the commercial process is much different than the homemade version. There are actually several ways that they can be made, but all methods depend on creating a cream paste by mixing powdered sugar with fat, a protein for aeration, and peppermint flavor, of course. The ingredient list for York and Pearson's version of the peppermint patty vary slightly, but both have those same common elements.

One manufacturing method uses extrusion technology. After the cream paste is made to the proper consistency, it's pressed through an extruder. In this case, the extruder simply forces the cream paste through a hole with the diameter of the patty. As the stiff cream comes through the hole, some sort of cutter, like a wire blade, cuts the candy into the proper thickness. Candy cream disks continuously plop onto a conveyor, one after another, after which they're chilled briefly to harden. The hardened patty is then passed through an enrober to coat the disk with chocolate. After passing through the cooling tunnel to solidify the chocolate, the product is ready for packaging.

Another manufacturing method uses single-shot depositing technology. This involves a complicated dual nozzle that shoots out both a liquid chocolate coating and a soft cream center at the same time. Coordination of the sequence of starting and stopping each flow is critical to the finished product. The chocolate flow must start milliseconds before the cream center starts to flow and shut off milliseconds after the cream stops to produce a cream disk perfectly enveloped in a layer of chocolate. The flow characteristics of cream and chocolate must be perfectly balanced. The end result when it works well is a very efficient and uniform operation.

Another similar product is the Haviland Thin Mint, a product of NECCO, which is made in yet another way. In this case, a thin creamy filling is deposited onto a conveyor; the fluid filling flows a little to form the patty shape. After the patty solidifies in a cooling tunnel, it goes through an enrober to provide the chocolate coating.

The taste and texture of these different peppermint patty products vary with process and formulation. Some are softer, some firmer, but do any really have the "snap" attributed to the original York product? Maybe not. In fact, the York Peppermint Patty currently contains an ingredient, invert sugar, that acts to soften the cream center. Invert sugar, a mixture of glucose and fructose (see Chap. 12), decreases the amount of crystalline sucrose in the cream center, making it less snappy.

Both the Pearson's patty and the Havilland mint also contain an ingredient that softens the center—invertase. This enzyme breaks down sucrose to create invert sugar, again softening the cream. The enzyme takes a little time to work, a useful trait in this case. The firm cream center made initially is easier to enrobe in chocolate, but then over the first few weeks of storage, the enzyme goes to work on softening the cream. The enzyme eventually stops acting once the invert sugar level has increased sufficiently, called product inhibition of the enzyme. The result, just like in Junior Mints (see Chap. 25), is a soft-ish cream center completely coated in chocolate.

Although the York Peppermint Pattie was developed prior (1940) to the introduction of the Peanuts character, Peppermint Pattie (1966), Charles Shulz, the creator of Peanuts, claimed that Peppermint Patty was named after a bowl of peppermints in his office, not the ones made in York. In fact, the York candy was only sold regionally until 1975, when the Peter-Paul company (of Mounds and Almond Joy fame) acquired the York brand and started national distribution. So it's quite likely that he never knew of the York candy before naming Patty.

25

Junior Mints

Cheek dimples make people more attractive. "Ain't she (or he) cute?" is a common expression when someone with dimples comes into view. Further, a dimple on only one cheek is especially rare, and makes people even more attractive. With a dimple on only one side, that would make Junior Mints the cutest candy around.

Where do dimples come from? On face cheeks, it's related to the nature of one of the cheek muscles, the zygomaticus major. Dimples are an inherited trait, actually a dominant trait (although, somewhat contradictorily, it's their relative rarity that makes people with them attractive). In Junior Mints, the dimple on one side arises from a physical phenomenon related to how they're made.

Junior Mints were developed by James Welch in 1949 at his candy factory in Cambridge, MA. They were named after his favorite Broadway play, Junior Miss, which at the time was also a movie and radio show. The choice of name provided an advertising bonanza.

As Kramer said in a Seinfeld episode, "Who's gonna turn down a Junior Mint? It's chocolate, it's peppermint—it's delicious! It's very refreshing!" And as later proven in the episode, they have magical curative powers when taken internally.

Curative powers notwithstanding, Junior Mints are delicious, providing a delectable contrast between the chocolate shell and the soft creamy mint inside.

The minty filling is made by mixing sugar and corn syrup with a whipped sugar syrup, called frappé, to lighten the texture. First, the sugar and corn syrup mixture is cooked to the appropriate temperature and water content (see Chap. 8). It's cooled quickly without

R.W. Hartel and AK. Hartel, *Candy Bites*, DOI 10.1007/978-1-4614-9383-9_25,
© Springer Science+Business Media New York 2014

agitation to just the right temperature, where it's intensely beaten to promote rapid sugar crystallization. The result is a semi-fluid and partially crystalline mass. At this point, it's blended with frappé to reduce density and give that desirable creamy texture.

Frappé is much like marshmallow, with corn syrup and sugar for bulk and protein to stabilize air bubbles. As the sugar syrup is whipped, the proteins coat the surfaces of the newly-formed air bubbles and prevent them from collapsing. Egg proteins, soy proteins and/or gelatin may be used to stabilize the air bubbles. In Junior Mints, the combination of gelatin and egg protein is used.

The mixture of cooked sugar and frappé, still somewhat liquid in nature, is deposited into depressions in a rubber conveyor that mold the candy into the shape of half a sphere. The conveyor carries the candy through a cooling tunnel where the sugar mass sets up into a firm piece as sugar crystallization continues. It is the high crystal content that gives the Junior Mint centers their firmness.

Those centers need to be firm because they are then coated with chocolate while tumbling in a panner, a revolving bowl (sort of like a stone polisher). The solidified Junior Mint centers tumble as the pan rotates and liquid chocolate is sprayed on the surface. The tumbling action smooths the liquid chocolate over the surface of the candy piece and the cold air in the pan causes the chocolate to solidify. Several sequential coatings of chocolate are applied to build the desired thickness of chocolate, a process called chocolate panning.

When the chocolate coating reaches the proper thickness, the pieces are removed from the pan and allowed to sit overnight so the cocoa butter in the chocolate can completely solidify. Then it's back into the pan for the polishing layer. Confectioners glaze, or edible shellac, is applied to the surface of the chocolate to provide the shiny appearance we value in a Junior Mint. The polish also prevents them from scuffing in the package during shipping and distribution. A similar process is used to polish various candies, from jelly beans to M&Ms.

In order to apply the chocolate layer during panning, the mint cream centers need to be sufficiently hard to stand up to the forces applied during tumbling. But the mint center of a Junior Mint is soft and creamy like a smooth fudge, not firm and hard. The secret is in the ingredient, invertase.

Invertase is an enzyme that breaks down sucrose, a disaccharide, into its component monosaccharides, fructose and glucose. As the sucrose is hydrolyzed into its component parts, invert sugar, the hydrolysis break-down product of sucrose (see Chap. 24) is produced. This causes the amount of sugar crystals to decrease. Both the invert sugar and reduced crystal content cause softening of the cream candy center. And it happens while the product is packaged and being shipped for sale. Although the chemistry is fairly complex, the end result is a softer cream center coated with chocolate.

The same trick is used to create the gooey cream center in Cordial Cherries. In the case of Junior Mints, the effect of the invertase is much less than in Cordial Cherries, with the cream center just turning soft, without becoming completely liquefied.

Junior Mints come in numerous varieties these days, although none have nearly the following as the original. There's the Junior Mints Deluxe, the jumbo-sized version. There's Junior Mint Mini's, the scaled-down version perfect for snacking. And, Junior Mints Inside Outs turn things around, with a dark chocolate center surrounded by a smooth, white peppermint candy coating.

So what gives the original Junior Mint the dimple on one side? It comes from the solidification process. When the cream is deposited into the rubber mold, it's spherical on the bottom side but flat on the top as the fluid cream flows to find it's own level (a liquid property). But then in the cooling process, crystallization of the sugars causes contraction of the cream, leading to formation of the dimple. Crystallization causes the sugar mass to contract since the density of the crystal form is higher than that of the liquid form. The result is the concavity, or dimple, on the top side of the cream center, which retains its shape even when coated with chocolate.

Ain't it cute?

26

National Candy Corn Day

Candy corn—these multi-colored kernels are the hallmark of Halloween. In fact, perhaps fittingly, National Candy Corn Day is the day before Halloween—October 30. The yellow bottom, orange middle and white tip of candy corn represent the colors of fall and harvest time, making them an icon of the season and a staple at Halloween.

It seems that candy corn is the one candy that generates the most polar responses—people either love them or hate them. For some, the waxy texture is unappealing, but it's the flavor that generates the most reaction, both pro and con. Some people love the sweetness while others complain that they're too sweet, without any real flavor.

Actually, candy corn has a unique flavor. It's not corn, even though there's plenty of corn used to make them, primarily in the form of corn syrup and even corn starch. Candy corn recipes call for butter, vanilla, and sometimes honey. This unique vanilla-butter-honey flavor is what gives candy corn kernels their distinct taste.

According to the National Confectioners Association, candy corn was developed in the 1880s at the Wunderle Candy Company in Philadelphia. The Wunderle Candy Company was bought by the Heide Candy Company, which, typical of the candy business (see Chap. 4), was then bought by another company and so on, until no remnants of the original company remain. But don't worry about finding them—over 35 million pounds of candy corn kernels are harvested every year. For those that like statistics (and have the time for such calculations), that's enough to go around the moon nearly 21 times.

R.W. Hartel and AK. Hartel, *Candy Bites*, DOI 10.1007/978-1-4614-9383-9_26,

Candy corn is what the candy maker calls a mallow cream (sometimes written mellowcrème). Crystallized sugar fondant is combined with a marshmallow-like ingredient called frappé to produce a tender candy with a clean bite. The main ingredients in mallow creams are sugar, corn syrup, colors and flavors, along with a whipping agent to hold the air bubbles in the frappé.

The starting point for candy corn is a highly crystallized sugar product called fondant (see Chap. 24). To make fondant, a sugar and corn syrup mixture is first heated to boil off water. The concentrated sugar syrup is then worked in a high-intensity mixer to promote formation of the numerous small sugar crystals found in fondant. Fondant is rarely eaten by itself—it's most often used as sugar crystal seeds in fudge or to make creams.

To make mallow creams, fondant is warmed slightly and diluted somewhat with a sugar syrup. Frappé, with its lower specific gravity, is then added to produce a lighter texture. Typically gelatin or egg protein is used in frappé to stabilize air bubbles, but soy protein can be used as well. The protein molecules form a protective layer around each air bubble, preventing the bubbles from coalescing.

The warm, creamy mixture of fondant and frappé is formed into candy corn kernels in a process known as starch molding. Wooden or fiberglass boards are filled with dried cornstarch powder—a little mineral oil allows the cornstarch to better hold shapes. The cornstarch surface is smoothed evenly and then candy-shaped depressions are pressed into the starch with a print board. For candy corn, the print board forms are triangular in shape, with the pointy end pressed down into the starch. The candy cream mixture deposited into the depressions takes the shape of the triangular mold, leaving one flat surface at the top, the bottom of the candy piece.

For candy corn, three sequential deposits are required; first the white tip is deposited, then the orange middle and finally, the yellow bottom. Timing is crucial to get the three layers to bond well together. If depositing is done too quickly, the colors are prone to smear, but if too much time elapses between each color, the layers won't bond together very well. In the 1880s and early 1900s, this process was done by hand. Now, it's done automatically in what's known as a starch mogul (see Chap. 36). A starch mogul automatically fills the boards with starch, forms the depression, deposits the right amount of the right color of cream into each depression, and stacks the boards for curing.

After being filled with candy, the trays are allowed to sit for a while as the cream sets, the sugar crystals finish growing, and moisture from the candy cream migrates into the cornstarch. This curing step allows the candy corn to develop a tender texture with a clean bite. The next day, the trays are upended and the candy corn separated from the cornstarch on a screen.

The final step is putting on the shine. Candy corn pieces are tumbled in a rotating pan and polishing agents applied. As the pan turns, the polishes are smoothed over the surface, leaving a nice shiny candy corn ready to eat.

Is there a "right" way to eat candy corn? A National Confectioners Association survey asked people how they preferred to eat candy corn. If you start by nibbling off the large yellow end, you're in a minority (10.6 percent). Most (46.8 percent) preferred to just pop the whole piece into the mouth rather than eat the narrow white end first (42.7 percent).

Candy pumpkins (and various other shapes like the scary jack o'lantern) are another version of a mallow cream candy. The same process is used to make candy pumpkins. Again, a press board is pushed into the corn starch to create a pumpkin shaped hole and, similar to candy corn, the fluid candy is deposited, but this time in two shots. The first shot fills the stem with green candy and the second shot fills in the orange pumpkin.

For those who remember, Chocolate Babies were another mallow cream candy popular decades ago. These were made in the same way, by pouring the fluid candy mass into a starch mold shaped in the form of a baby. Flavored with cocoa, they had the texture of candy corn but with a chocolate flavor. We don't know who came up with the idea, but eating babies made from chocolate mallow cream seems a little odd.

Although traditional candy corn is yellow, orange and white to denote Fall colors, other varieties are available as well. "Indian" corn is available in the Fall with a brown base, "reindeer" corn arrives at Christmas with a red base, followed by a green middle and the white tip, and "cupid" corn is for Valentine, with red, pink and white.

What do you do with left-over candy corn, whatever the color? One funny t-shirt shows five or six different uses, from traffic cones to reverse hearing aids (ear plugs?). Or as camouflage for better candy? One astronaut found the coolest (for a scientist) use—demonstrating the principles of soap molecules in a droplet, but done in zero-gravity in space. Seriously, check it out by searching on Don Pettit, astronaut, for his candy corn in space experiment.

27

Maple Syrup Candies: A Natural Treat?

Who was it do you think that first figured out that sap from maple trees was sweet? Perhaps an ancestor who, in the early days of spring when trees dormant throughout the winter began to stir, was so hungry that he decided to try to eat the tree? After the sap started to run a little, he got some on his fingers and, like many of us would do, licked it off. To his surprise, it was slightly sweet and seeing as he was starving, he ate more.

These same ancestors then figured out how to concentrate the sweetness in the sap by cooking it on the fire. Some cavemen somewhere must have left a batch of overly concentrated maple syrup out so that the sugar crystallized. Instead of tossing it away, they tried eating it and found, to their amazement, that it was sweet and tasty. The first maple sugar candy was born.

That's as natural as it gets, right? But let's look at how these candies are made today and then decide if we still think it's natural.

First, exactly what's in maple syrup? Sucrose. Maple syrup can be anywhere between 88 and 99 percent sucrose. It also may contain up to 0 and 11 percent invert sugar, which is just a sugar-makers way of saying glucose and fructose, the by-products of sucrose break-down. Sucrose is a disaccharide, made up of a molecule of glucose and fructose bonded together. Under certain conditions, high temperature and acidic, the sucrose breaks down, or hydro-lyzes, into glucose and fructose, often called invert sugar (see Chap. 12).

Maple syrup also contains small amounts of other impurities such as organic acids (malic acid mostly) and minerals. Maple syrup is relatively high in potassium and calcium. And of course, there are

R.W. Hartel and AK. Hartel, *Candy Bites*, DOI 10.1007/978-1-4614-9383-9_27,
© Springer Science+Business Media New York 2014

essential flavor compounds that provide the unique flavor of maple syrup. These flavors, not present in raw sap, are developed during the cooking step.

Since maple syrup is mostly sugar, it's not a surprise that it can easily be made into candy. In fact, maple syrup candy is essentially a highly crystallized form of maple syrup. But to control the process to make the smoothest, most delectable confection, it's critical that the crystallization process is carefully controlled.

The process for making the candy starts with pure maple syrup, lots of it, by the barrelful. Preferably, the syrup contains relatively low levels (1.5–2 percent) of invert sugar. Too much invert sugar acts like a candy "doctor", inhibiting sucrose crystallization (see Chap. 12). Since crystallization is desired in this candy, we need to minimize the amount of invert sugar to maximize the amount of crystals formed.

To make maple sugar candy, maple syrup is split into three allotments, each of which goes through a different process until, at the end they're combined into the sparkly sweet candy.

The first allotment of syrup is used to make a maple syrup fondant. In a cooking kettle, the maple syrup (about 32 percent water initially) is concentrated to about 10–12 percent water by boiling to 244 °F. This concentrates the sugar so it will be easy to crystallize, although it's very important that it doesn't crystallize yet. The hot concentrated syrup is poured onto a cold table and cooled quickly and carefully to about 125–130 °F, with minimal agitation. If done right, the crystal-free concentrated syrup is highly supersaturated and intense agitation promotes rapid crystallization all at once. Getting all the sugar to crystallize out at the same time means we create billions and billions (a la Carl Sagan) of very small crystals that give a smooth, velvety texture. If the concentrated syrup crystallizes too soon, large crystals form and this results in a coarse candy texture.

Fondant is essentially a highly crystalline confection typically used to make cream candies or, in this case, maple syrup candy. Because of the high (about 50–60 percent) crystal content, fondant is somewhat firm and solid. Ever had a cordial cherry that had a firm center? Or the center of a Junior Mint. That's the texture of fondant.

A second allotment of syrup is concentrated to provide a thinning syrup for the fondant. This syrup is also cooked to about 244 °F to reduce its water content to 10–12 percent. After cooling carefully to about 180 °F to avoid crystal formation, a set amount of fondant is added to the thinning syrup. At these elevated temperatures, the fondant is dispersed with some of the sugar crystals dissolving. If done correctly, this results in a thinner fluid-like material that still contains sufficient sucrose crystals for the next step. Both temperature and water content are critical to this step.

The fluid candy mass is then poured or deposited into molds of whatever shape is desired. Rubber molds in the shape of a maple leaf or a holiday Santa are filled with the fluid candy mass and then allowed to slowly cool. During cooling, the sugar crystals in the fluid mass crystallize out and solidify the piece to give the desired texture of maple sugar candies. Once cooled, the candies are simply popped out of the rubber mold, ready for the next step. Carefully controlling every step in the process is necessary to get the smooth, creamy candy without white spotting.

But it's not done yet. If we stopped here, we would have a delectable candy, but it wouldn't last very long. Any humidity or heat would cause the candy to spoil very quickly. Specifically, moisture in the air would enter the candy, cause sugar crystals to dissolve and leave a syrupy residue. This unsightly separation can be prevented by putting a protective layer of crystals on the candy surface.

The third allotment of syrup is prepared for the sugaring step, where small crystals are created on the surface of the candy as a protective coat. This syrup allotment is again heated to drive off a bit of water, but not nearly as concentrated as the other two allotments. The aim is to create a slightly supersaturated sugar syrup that allows crystals to grow on an already crystalline candy but does not crystallize itself. Baskets filled with the uncoated maple sugar candies are lowered into the crystallizer syrup and allowed to settle there for several hours. During the time the candies are immersed in the supersaturated syrup, small sugar crystals grow on the surface. Careful control of temperature and concentration is important so that these crystals don't grow too large.

Once the surface crystals have formed, the buckets of candy are raised and any remaining syrup is allowed to drain. If done correctly, the numerous small crystals on the surface of the candy give that attractive sparkle to the maple sugar candy. But more importantly, those crystals act as a barrier to humidity in the air and allow an extended shelf life.

So, production of maple sugar candies involves simple evaporation, crystallizing and drying steps. No chemicals or additives are needed. In fact, the label simply says the candy contains only one ingredient—maple syrup. The difference between the original maple syrup and a maple sugar candy is simply one of form. In the syrup, the sugar is liquid; in the candy, a portion of the sugar has been crystallized. It's really no different than letting the sugar crystallize naturally in maple syrup as water evaporates, other than we've controlled the manner of that crystallization to give a smooth, creamy candy.

Some may argue that we've still "processed" the maple syrup so the candy is no longer natural. But let's walk through the steps that go from maple tree sap to maple sugar candy and talk about the changes that take place. Is maple syrup really natural?

In late winter and early spring, the sap that runs in the xylem of the sugar maple tree is collected by tapping holes in the tree. Sap, the raw material of maple sugar candy, is the lifeblood of a tree. It contains water and nutrients, mostly sugars and minerals. It's sticky and not very tasty. It does not yet have a maple flavor—it's definitely not something that you'd want to eat unless you're starving. Yet, it's raw and natural.

When cooked, maple sap is magically transformed into a sweet, flavorful syrup. In part, this is from the concentration of the sugar but it's also in part from the chemical transformations due to the Maillard browning reaction. Sugars and proteins (present in very minor amounts in maple sap) react to produce the maple flavors and brown colors of maple syrup. Note that maple syrup would definitely not satisfy a raw food aficionado because of the high temperatures (215–220 °F) needed to develop the desired color and flavors. Still, most of us consider maple syrup to be a natural food because of the limited amount of processing involved.

As we've seen above, the steps needed to convert maple syrup into maple sugar candy also only involve cooking and cooling (to allow crystallization). In a very real sense, the only ingredient used is natural; we've added nothing else and only removed water. The only real difference between the two is the state of the sugar molecules—liquid or crystal. In this sense though, maple syrup may be considered more natural than maple sugar candy because the sugar molecules are in the same state as found in the raw material—the liquid form.

Is maple sugar candy natural or not? The point is that, in some cases, it's difficult to decide where the term "natural" ends and "processed" begins.

28

Caramel: Controlled Scorching of Milk?

I once heard it said that making caramel involved the controlled scorching of milk—an interesting concept, but not always correct. Sure, one method of making caramel involves heating condensed milk and sugar through boiling to generate caramel flavors and brown color. In a sense, this is controlled scorching of milk. But caramels can also be made by scorching the sugar, not the condensed milk.

Being used to making caramel with commercial methods, by scorching milk so to speak, I was a little taken aback one day when a chef friend of mine provided instructions for making a caramel filling for chocolate. She instructed me to first heat the granulated sugar and corn syrup in a pan until it reached the desired color. She didn't supply a temperature, which, as a scientist looking for specific targets, confused me to no end. She said to just keep cooking it until it reached that visual color endpoint. When pressed, she guessed the temperature might be somewhere around 350–360 °F but couldn't be any more specific. Her instructions continued to then add cream or condensed milk to that scorched sugar (which cooled it considerably), and then finally, to boil that mixture briefly until it reached the set temperature of 244 °F (see Chap. 8).

Her chef approach required the sugars to be scorched first to generate the caramel flavor and brown color, whereas my more commercial approach required the sugar and milk component to be cooked together until it reached 244 °F. Since these are different approaches, it made sense to think that our caramels would taste different. But after careful comparison, making sure the exact same ingredients and levels were used in both processes, we were unable

R.W. Hartel and AK. Hartel, *Candy Bites*, DOI 10.1007/978-1-4614-9383-9_28,   111
© Springer Science+Business Media New York 2014

to differentiate their flavors. Maybe we weren't sensitive enough to pick up the different nuances in flavor and aroma between the two versions. Or maybe understanding the chemistry can help explain this apparent contradiction.

The heat involved in the process of making caramel, as with many cooking steps, initiates various chemical reactions that cause changes in the components. In this case, they lead to the final colors and flavors. In caramels, there are two important types of reactions, both called browning reactions, with their relative importance depending on how the caramel is made.

The first is Maillard browning. Named after a French chemist, Maillard browning is a reaction between certain types of sugars and proteins. Well, it's really a series of reactions that starts between a reducing sugar (glucose, fructose, lactose, but not sucrose) and a protein. The exact path of the complex series of steps depends on many parameters, including the ingredients present, pH, and temperature. The nature of the flavors, aromas and color compounds produced depends on each of those parameters, and more.

As the reaction proceeds, a variety of compounds are produced. Getting slightly technical, they include pyrazines, pyrroles, pyridines, pyranones, oxazoles, oxalines, furans, and furanones, compounds that are volatile so readily escape into the air (or our nasal passage) to be detected as aromas. They contribute various characteristics, including caramel-like, cooked, roasted, sweet, burnt, pungent and nutty notes. Also, through the reaction process, highly reactive cyclic compounds are produced, which rapidly polymerize to form melanoidins, the colorant components of the reaction.

The other browning reaction of importance in making caramels is caramelization. This is simply a reaction of reducing sugars when exposed to sufficiently high temperatures—no proteins are needed. As with Maillard browning, caramelization is a complex series of reactions with the specific colors, flavors and aromas generated depending on the ingredients and the processing conditions. In caramelization, the initiation step is dehydration of a reducing sugar caused by the high temperature. From here on, the reaction path is quite similar to that of Maillard browning and so, many of the same flavor and aroma compounds are produced. The melanoidins are similar polymeric compounds (although without any protein residues).

My guess, based on the fact that we call it caramel, is that the first caramels produced must have been made by scorching sugars, through caramelization. Hence, the name caramel. But one might ask, which came first, the caramel or the reaction, like the chicken and the egg?

Controlled caramelization of sugars is also used to generate caramel color and caramel flavor, both used in a variety of foods. Caramel color is used in colas and some spirits, for example, while caramel flavor finds application in licorice, ice cream, and even caramel latte. There's even a differentiation of caramel color, depending on the temperatures to which the sugars are heated and what other ingredients are added. Higher temperatures give a darker brown color with more pungent flavor notes.

Technically, sucrose doesn't undergo Maillard browning or caramelization because it's not a reducing sugar. Sucrose is a disaccharide made up of two reducing sugars, glucose and fructose, linked in such a way that it doesn't have a reducing carbon. Before sucrose can be caramelized, it has to go through a preliminary reaction—hydrolysis. The heat causes the bond between the glucose and fructose to be broken, or hydrolyzed, creating one molecule each of glucose and fructose. These breakdown products are then available to initiate the caramelization reaction.

Can you list some other foods where either Maillard browning or caramelization is important? Actually, these browning reactions are key in a wide range of foods. Roasting cocoa beans into chocolate or coffee beans into coffee both involve Maillard browning and caramelization, as does turning grapes into raisins. Browning of bread in an oven or toaster utilizes the Maillard reaction. Even cooking meat involves the Maillard browning reaction. The different flavors arise because of the different sugars and proteins present in each food that participate in the reaction.

Although chefs and commercial caramel makers use different approaches and rely on different reactions to create caramel flavor and aroma, the end result of each is quite similar—a tasty caramel.

29

A Caramel Family

What does a candy caramel family look like? A Sugar Daddy with a Sugar Mama, with several Sugar Babies and maybe a Junior Caramel or two. This happy family has gone through some rough times together, though, and experienced its share of hardships and change. From a candy science standpoint, they also demonstrate the variety of caramels found on the market.

Sugar Daddies were the first to arrive, appearing in 1925 with the original name of Papa Sucker. The name was changed to Sugar Daddy in 1932, perhaps to play on the growing popularity of rich old men taking younger women under their wing during the Great Depression. Sugar Babies came along in 1935 as a spin-off of Sugar Daddies, supposedly named after those young women being taken care of by the sugar daddies. The first iteration of the Sugar Mama caramel didn't come out until 1965 and Junior Caramels are also a relatively recent addition to the family line, adopted from an earlier candy.

A Sugar Daddy is a hard chewy caramel, so hard that it stands by itself on a stick, like a caramel lollipop. You have to be patient to eat a Sugar Daddy. You suck on it, you warm it up, you stretch it out, and eventually you can start gnawing away at the edges. You tease it with your tongue into something you can eventually chew. It's good movie theater candy—if you tease it right, it can last longer than most movies.

Sugar Babies are little spheres of really chewy caramel, nowhere near as hard as the Sugar Daddy. The chewy texture is controlled through the finished water content, with Sugar Babies having slightly higher water content than Sugar Daddies (see Chap. 30).

R.W. Hartel and AK. Hartel, *Candy Bites*, DOI 10.1007/978-1-4614-9383-9_29,
© Springer Science+Business Media New York 2014

To make them truly distinct, the Sugar Baby caramel spheres are coated with coarse granular, caramelized sugar crystals to provide contrasting texture. Sugar Babies also make great movie candy—a box of these babies can last through an entire movie. Now there's even a chocolate covered Sugar Baby on the market—chewy caramel coated in real milk chocolate. Sounds like competition for another tooth-sticking caramel candy, the Milk Dud.

Junior Caramels may be the brown sheep of the family, or at least they're brown-coated. They're an easy-to-bite caramel sphere coated in chocolate. The caramel is highly grained (lots of small sugar crystals) to give a softer, less chewy bite (see Chap. 33) than the chocolate-covered Sugar Babies noted above. Junior Caramels still make a good movie candy, but don't wreak as much havoc on filling-filled teeth.

The only thing missing from this perfect caramel family is the Sugar Mama. The original 1965 version of the Sugar Mama didn't last very long. It was a hard caramel on a stick, the same as the Sugar Daddy, but coated with chocolate. A great idea, but it never caught on. Way before the caramel started to soften and become edible, the chocolate was already gone. Chocolate and caramel make a great match; it's just that in this case the match wasn't made in heaven.

People didn't buy it—no sales, no profit. The original Sugar Mama was let go from the caramel family.

Although the original Sugar Mama hit the dead candy list in the mid 1980s, the dream of a functional sugar caramel family continued. In the mid-2000s, a new Sugar Mama made her appearance on the market, once more completing the family. The new Mama was a rectangular chunk of individually-wrapped caramel that was easy to bite through, like the center of a Junior Caramel. Unfortunately, this Mama didn't last long either. Within just a few years it was already on the dead candy list.

Why have Sugar Babies and Sugar Daddies lasted so long while the original Sugar Mama hardly made it through her teens and the second was gone faster than that? It's business—companies make decisions all the time to drop impoverished product lines. But is it that the product isn't any good or it just wasn't marketed well enough? One side of the coin says an excellent product sells itself even without advertising. On the other side, marketing people say they can sell anything, no matter what the quality of the product.

Without marketing, how can a new candy product possibly succeed? Sugar Babies and Sugar Daddies have succeeded over the years with limited marketing mostly due to their long-standing reputations. Undoubtedly, at one time, the manufacturer put money into marketing them. Now they're so entrenched in our culture that it would be hard to imagine them being dropped (although other once-popular brands are now gone, so it's not inconceivable).

Sometimes, a simple change can make all the difference. An interesting example of remarketing a candy to boost sales is the Junior Caramel, which originally started as the Pom-Pom. The name was switched to play off another successful product—the Junior Mint. By connecting a less successful brand to a candy powerhouse, this simple name change gave new life to the Junior Caramel.

A new candy, like a Sugar Mama, is easily lost in the candy aisle, where hundreds of candies call out for our attention. Unless directed there by a clever marketing campaign, why would we choose Sugar Mamas over any other candy?

Launching a completely new product is a challenge these days, especially in the highly competitive candy aisle. The key to a successful new product launch is a good marketing campaign to get people to try it. Perhaps if the Sugar Mama was located right between the Sugar Daddy and Sugar Babies on the candy shelf, consumers would be intrigued and try it. Then, the product would have to be good enough that people would come back to it again and again. Marketing can help repeat business, sometimes making the difference between success and failure of a new product.

Perhaps the Tootsie Roll Company, manufacturers of the caramel family, will develop a third version of the Sugar Mama. If it's going to be successful, though, they'll either have to somehow make it so compelling for us to buy, perhaps by spending some marketing dollars to get and retain our business. Only time will tell.

30

Caramel Cold Flow

To a caramel maker, the term "cold flow" signals the end of shelf life, the end of useful quality, the point at which a consumer would say "This ain't no good." There's lots of reasons why a consumer would not purchase a candy; one of them is shape, although it's much further down the list than mold, bloom, and yucky appearance. A consumer expects a piece with a certain shape, and when that shape changes due to cold flow, consumers think "Something's wrong with that caramel."

Interestingly, the dictionary definition of cold flow is "the viscous flow of a solid at room temperature". Wait, viscous flow of a solid? What does that mean? Let's look first at what cold flow means to a candy maker and then come back to this apparent contradiction.

Cold flow to the caramel maker is defined as the room temperature collapse of a perfect piece of candy. A gourmet caramel maker might present her caramel in the form of a wax paper-wrapped log. When fresh, the rolled log forms a perfect sphere viewed end-on. When cold flow occurs, the spherical log, and remember it's sitting at room temperature, collapses due to the force of gravity pulling it down. At first, the cross section becomes slightly oval and then becomes increasingly oblate as cold flow continues. Eventually, given sufficient time, the caramel flows completely into a flattened caramel pancake. As the tar pitch experiments show (see Chap. 9), even solids can "flow before the Lord." Caramel is another good example of a "soft solid."

To understand cold flow, and the dictionary definition, we need to understand caramel. Caramel is a soft solid, a term that

R.W. Hartel and AK. Hartel, *Candy Bites*, DOI 10.1007/978-1-4614-9383-9_30, © Springer Science+Business Media New York 2014

recognizes materials that have both solid-like and fluid-like characteristics. While the candy maker wants the caramel to act more solid-like, to resist cold flow, sometimes it acts more fluid-like and collapses in cold flow.

What gives caramel its solid-like properties? It's structures. Sure, caramel seems like a pretty uniform mixture of ingredients. Cut a caramel in bits and look at the cross section—to the eye, it looks the same all the way around. But at a microscopic level, caramel has a lot of structure, and it's this structure that protects it from the force of gravity.

Most of the caramel is an aqueous solution of sugar, corn syrup, milk proteins, and other milk ingredients (lactose, minerals). This aqueous solution, called the continuous phase, has fluid-like properties that depend on water content. Caramel with high water content is less viscous than one with low water content. Water is the first element the candy maker has to combat cold flow.

To completely prevent cold flow, the caramel maker could boil off enough water so that this continuous phase has such a high viscosity that it never flows. The caramel hard candy and even the caramel on a stick are examples of this. In general, higher water content means softer, more fluid caramel.

Sometimes a soft caramel is desired, since a softer caramel is easier to eat. A gooey caramel for inside a chocolate shell would also have higher water content. At the extreme, caramel sauce has even higher water content and is designed to flow (tastes great over some ice cream with whipped cream and a cherry on top).

The caramel where cold flow is a problem is the one with intermediate moisture content that's designed to be chewy yet firm enough to resist flow. That's where the rest of the caramel microstructures come into play. There are several elements that provide structure to prevent cold flow.

For one, caramel is an emulsion. The fat in the caramel is dispersed as small droplets throughout the amorphous matrix. Furthermore, these droplets contain fat crystals so they're not fluid themselves. Think of butter—at room temperature it's solid enough to retain its shape without collapsing. The milk fat in butter at room temperature is partially crystalline (and thus, also partially liquid). The presence of numerous partially-crystalline fat globules provides a resistance to prevent the amorphous continuous phase from flowing. In general, more fat and the more solid the fat, the less cold flow.

To help even more, the milk proteins aggregate during caramel cooking and provide a second network around the fat globules that further helps to prevent cold flow. Most caramel makers know that how the milk ingredient is treated prior to making caramel plays a big role in the final product, particularly as to whether cold flow occurs or not. The pasteurization step that goes into milk processing and the evaporation step in making condensed milk both influence the nature of the protein. Some caramel makers say that pre-heating milk helps control the protein to avert cold flow.

Increasing protein level helps to combat cold flow, but it only works up to a point. When protein level is too high, protein graining becomes a danger. This is when the protein aggregates into chunks that are large enough to see, giving the caramel mass a tapioca-like texture. These chunks prevent cold flow, but that's the only good thing about it. When protein grain occurs, there's nothing more you can do except toss out the batch.

Usually, the fat and protein structures are sufficient to prevent cold flow, but not always. Another tool the caramel maker has to ward off cold flow is to add some sugar crystals. Like the semi-solid fat globules, these hard bits of sugar crystals work well to support the amorphous phase and prevent flow. The only problem is that sugar crystals change the texture, making the caramel less chewy. In fact, if there are too many sugar crystals, caramel becomes fudge, a softer, sometimes brittle candy without the chewy characteristics of caramel.

As usual, what seems like a perfectly simple candy—in this case, caramel—turns out to be quite a complex material, whose characteristics can change with both the ingredients used and the manufacturing properties. Caramel properties can be controlled by the candy maker to achieve a wide range of characteristics.

31

Tootsie Roll Pops

Probably the biggest, but not the only, mystery surrounding the Tootsie Roll Pop is how many licks it takes to eat one. The wise owl advertisements from the 1970s used to play on that theme, although there never was a good answer. As the owl suggests, most people can't simply lick their way through one; they end up crunching it at some point to get to the chocolatey center.

The Tootsie Roll, first developed in 1896 and named after the daughter of the inventor, is an interesting candy. It's not really a caramel or nougat or a chew, yet it has characteristics of each. The company probably likes it that we don't have a good way to characterize Tootsie Rolls—it's in a class all its own.

But to the candy maker, the Tootsie Roll falls primarily in the caramel family of candies, although with a unique twist. As seen earlier (see Chap. 28) what makes caramel unique among candies is that it contains sugars and a milk ingredient, with the color and flavor resulting from browning reactions (Maillard browning or caramelization). Look at the ingredient list of a Tootsie Roll and most of the same ingredients appear as used in caramel. What makes the Tootsie Roll unique though is the cocoa flavoring.

As we learned earlier, we control the scorching of milk (or sugar) to generate the caramel flavor and color. But suppose we simply cook that sugar-dairy mixture really fast so there isn't time for much browning to occur? And then add cocoa powder to provide a unique chocolatey flavor? Voila—the Tootsie Roll—a "white" caramel, with no caramel flavor or brown color, flavored with cocoa. Or it can be flavored with anything, from vanilla to strawberry, options provided us by Tootsie Roll as well.

R.W. Hartel and AK. Hartel, *Candy Bites*, DOI 10.1007/978-1-4614-9383-9_31,

One other unique attribute of a Tootsie Roll is its semi-chewy texture. It sticks to the teeth a little, but not nearly as bad as a really chewy caramel. The Tootsie Roll contains numerous small sugar crystals; the company specifically creates these in the process through intense agitation after cooking.

Look at a Tootsie Roll under the microscope and see all the small sugar crystals. If you're interested in doing this yourself, simply take a thin slice of the candy from the interior using a razor blade. Place the thin slice on a microscope slide and place a cover slip on top. Smear the candy into an even thinner layer by moving the cover slip back and forth using a rubber-tipped tweezer (or some similar utensil). If the layer is too thick, a few drops of a dispersing agent like acetone (nail polish remover) might help thin it out. Under the microscope, you should be able to identify small particles (about 10–15 μm in size, or about a tenth as large as a human hair). A polarized light microscope would help clearly bring them into view if you have one.

These small crystals serve a purpose, to create a somewhat "short" texture. Compared to a really chewy caramel, like the center of a Milk Dud, Tootsie Rolls have short texture due to the presence of the numerous small crystals. Pull one apart and it breaks relatively clean (compared to the caramel of a Milk Dud, which will stretch out a long way before the strand breaks). The sugar crystals in a Tootsie Roll break up the amorphous phase of sugars and protein a little, but because there are relatively fewer of them, they don't give a fudgy texture.

Tootsie Rolls are good to eat plain. In fact, 64 million Tootsie Rolls are made each day to keep up with our demand. But, it doesn't end there. Tootsie Roll Pops add extra excitement to the standard Tootsie Roll.

What's your favorite flavor of Tootsie Roll Pop? Chocolate? Grape? One survey shows that the most popular flavors were cherry (53 percent) and raspberry (40 percent), with chocolate (20 percent) and grape (13 percent) falling far behind. Sometimes they come out with new flavors—pomegranate or banana?

A Tootsie Roll Pop is simply a Tootsie Roll on a stick coated with hard candy. The manufacturing process generally follows the same process used by all filled lollipops, whether filled with gum or something else. The hard candy and Tootsie Roll portions are made in separate batches and brought together in the batch roller. This device is made of two tapered cones that rotate in opposite directions, with the candy mass being formed into a rope between the rollers.

The warm hard candy shell rotates on the batch roller while the hot Tootsie Roll mass, still in a fluid-like state, is fed into the filling feeder. A thin rope of Tootsie Roll coated with a shell of hard candy comes off the batch roller and sized to the proper dimension for the Pop. A cutting device stamps out each piece while the candy rope is still malleable and a stick is inserted. The lollipop then cools quickly to room temperature to set the hard candy into a sugar glass. The Tootsie Roll solidifies at the same time.

Urban legend has it that if your Tootsie Roll Pop wrapper shows an Indian shooting an arrow at a start, you'll get a free Tootsie Roll Pop. Lots of stories and blogs have been written about the subject, but apparently the Tootsie Roll company has disavowed this claim and does not actually send free Pops to kids that submit their wrappers. Instead, they're now supposedly sending a short story, more of a mythology, about how the Indian chief developed the Tootsie Roll Pop many years ago to differentiate his lollipop from all the others. Still, why would they even have that picture on the wrapper if there wasn't something behind it?

Back to the number of licks issue. Some interesting "scientific" studies have gone into determining the exact number. In fact, in separate studies, engineering students at both Purdue and Michigan developed automated licking machines to get to the bottom of the Tootsie Roll Pop. The Purdue group measured 364 licks by machine while the Michigan study concluded that 411 licks were needed. Interestingly, only 252 (on average) licks were required for human lickers. Perhaps saliva provides the difference between human and machine.

32

Cajeta

As we pull into the yard of Fat Toad Farm in central Vermont, we can hear the goats bleating off to the right, either in the barn or off in the field grazing. Dogs and chickens run free in the yard.

We're here to visit their cajeta production facility. Cajeta is a goat milk caramel used either as a sauce over ice cream, as a spread on toast or cookies, or as an ingredient in other recipes. Traditional Mexican cajeta is made with goat's milk, although some recipes call for a mixture of goat and cow, probably to mitigate the distinctive flavor of goat's milk.

To make cajeta, fresh goat's milk is sweetened with sugar and heated in a kettle. Usually, a small amount of baking soda is added too. Sometimes starch is added to enhance thickening. Flavors can be added either during the cook or afterwards to enhance the cajeta. Vanilla or coffee beans can be added during the cook to infuse the batch with flavor. Solid bits are filtered off before bottling. Alcohol can be added as well. If added as the cook is completed, at about 220 °F, the alcohol flashes off leaving the spirits flavor behind in the cajeta.

Since goat's milk is close to 90 percent water, it takes a long time to boil it down to caramel. The caramel-like color, flavors and aromas develop slowly during cooking through the Maillard browning reaction between proteins and reducing sugars (see Chap. 12). Milk contains plenty of each, although the sugar in milk is lactose.

The long cook time, on the order of five to six hours here at Fat Toad Farm, provides an opportunity for an upper body workout since the milk has to be constantly stirred to prevent scorching. Milk proteins are notorious for forming ugly black specks if not

R.W. Hartel and AK. Hartel, *Candy Bites*, DOI 10.1007/978-1-4614-9383-9_32,
© Springer Science+Business Media New York 2014

stirred adequately. There's even a proper way to stir caramel so the entire batch gets mixed uniformly without spilling over the sides. If you stir by just swirling the milk around in circles, it vortexes and flies up and over the side onto the stove or floor. A good figure eight motion is recommended to continually sweep the surface clear without making a mess. It also provides a nice "total-arm" workout, using a variety of muscles.

Dulce de leche is similar to cajeta, except it's usually made with cow's milk. Argentina claims to be the birthplace of dulce de leche. In fact, the word, cajeta, can be considered offensive there. An Argentinian friend said "Argentineans get mad because we believe that dulce de leche is an Argentinean invention and we are too proud to admit that the 'same' product exists in other countries." So be careful how you speak about cajeta or dulce de leche when you're in Argentina.

How is cajeta, or dulce de leche for that matter, different from commercial caramel? All commercial caramels start with a much more concentrated dairy ingredient to minimize the cooking time. Sweetened condensed milk is a common starting ingredient, where much of the water from milk has been evaporated off in large continuous evaporators (rather than in an open kettle on a stove) and replaced with sugar. Some cajeta or dulce de leche recipes suggest starting with sweetened condensed milk because it saves time, but the general sense is that it's not as good. It doesn't have the same fresh milk flavor.

Many people argue that cajeta and dulce de leche have a richer, creamier texture than regular caramel. There's a good reason for that—it's got more protein. A regular caramel, if any caramel could be defined as "regular," probably contains only 2.5–4 percent protein while cajeta, which uses more milk as an ingredient, has on the order of 7 percent. That higher protein content helps provide a creamier texture.

But the baking soda that's added to cajeta also has an effect on texture. In fact, if baking soda wasn't added to cajeta, that high protein level would be a problem. At such high levels, the proteins would tend to aggregate to form a sort of curd in the caramel, no matter how fast you stirred. The caramel maker calls this protein graining. When it occurs to an excessive level, the caramel takes on a tapioca-like texture and appearance. Under a microscope, large protein aggregates, or grains, can be observed.

The baking soda helps avert this problem, at least most of the time. By changing the pH of the caramel mix as it's cooking, the baking soda helps to keep the proteins from excessive aggregation. This also helps produce a creamier caramel; it modifies the flavors and aromas a little as well since pH is one of the governing factors of the Maillard browning reaction.

Interestingly, cajeta makers have found that goat's milk at certain times of the year is more prone to protein graining than at other times. Specifically, goat milk from late in the season, heading into the Fall months, is significantly more prone to graining than at other times, even with baking soda in the cajeta recipe. This is most likely because goat milk has higher protein content at that time of year, although it's possible that changing ratios of certain types of proteins may also play a role.

Why does milk change over the season? There are a couple of reasons. First, the nutritional needs of the calf change as they age and so mother's milk has adapted to that change accordingly. Analytical studies have found changes in sugar, proteins, minerals and fat during the different stages of lactation. Another factor is that the feed is different, especially if the cows or goats are grazing. Spring forage plants are different from Fall plants and this leads to changes in how the cow or goat converts the feed into milk, analogous to a breast-feeding mother whose milk varies slightly depending on what she eats.

Small cajeta companies, like Fat Toad Farms, must deal with these natural variations. They are continuously looking for methods cook their goat milk to prevent protein graining and other problems at different times of the year.

Cajeta or dulce de leche? At Fat Toad Farms, goats reign so cajeta is what they sell. But both are a wonderful, natural treat.

33

The Fudge Factor

"Oh fudge!" Supposedly that's what the cook said in an 1886 Baltimore kitchen when a batch of caramel went wrong. Although he used the term as a mild expletive, the name fudge stuck to describe the crystallized caramel confection produced that day.

Fudge is an interesting word. It's often used in various ways and situations.

Fudge can mean nonsense or humbug, as in "fudge on that." Fudge means to fake or falsify, as in "I fudged the data," a use perhaps derived from Captain "Lying" Fudge, an 1800s sea captain known for telling tall tales. Fudge sometimes is used to denote indecision, as in "he fudged on that issue," or in a variation used by engineers, the term denotes lack of certainty, as in "use a fudge factor."

To the engineer, that means calculating a value to 7 (or more) decimal places, but then adding another 50 percent to the calculated value to account for the fact that the numbers that went into the calculation weren't very accurate. Engineers are known "to calculate anything even when there isn't enough data." We then apply a fudge factor to make up for the fact that we often don't know things very accurately.

In the candy world, fudge is a delicious milk-based candy, often associated with tourist resorts. However, fudge is also a term found on packages to describe chocolate-like products, as in fudge coatings on cookies. Or it describes the chocolate syrup used to coat ice cream, as in hot fudge sundaes.

From a technical standpoint, the confection we know as fudge is crystallized, or grained, caramel, a milk-based candy. Fudge is formulated to contain more sucrose than corn syrup so that the sugar syrup is supersaturated and susceptible to crystallization. To the candy maker, crystallization is called graining. Either through agitation or seeding with fondant, graining is initiated in fudge to produce numerous small crystals that impart an extremely short texture. Fudge is soft and breaks easily because of the sugar crystals.

You can almost imagine the consternation of the first fudge maker who was really trying to make caramel. He ended up with crystals that gave a short texture instead of the chewy strands of caramel he expected. Short texture is measured "scientifically" as the distance you can pull a candy apart before the strand breaks. An ungrained caramel stretches a considerable distance before it breaks, whereas fudge breaks apart almost immediately. Try it at home. Slowly pull an ungrained caramel, like the center of a Milk Dud or a Riesen's chewy caramel, apart in your hands. It should stretch a long way. Now try it with a Tootsie Roll (a type of grained caramel). It has a much "shorter" stretch. And fudge should be even shorter.

Nowadays, fudge often connotes an artisan-style candy. In fact, some artisan fudge makers swear by their method of making fudge, keeping it secret—handed down from generation to generation. In their shops, you see the cooked mass poured out onto marble slabs and left briefly to cool. The goal is to find just the right temperature where swift mixing and agitation leads to spontaneous sugar crystallization. In this method of making fudge, the sugar crystals are created when the supersaturated sugar syrup is repeatedly mixed and sheared—the mechanical agitation promotes rapid nucleation, or the formation of numerous crystals.

Other fudge makers use a simpler approach, just adding fondant to seed the supersaturated sugar syrup to get crystals to form. That's what the fudge maker at the candy shop in Provincetown, MA did (see Chap. 1). Fondant is a highly crystallized candy base usually used for creating cream candies (see Chap. 31). Fondant contains about 50–60 percent of its mass in the form of minute, less than 10 μm, sugar crystals held together by a saturated sugar solution.

When fondant is added to the warm sugar syrup in a fudge recipe, some of the sugar crystals dissolve away. However, the crystals that remain act as seeds for graining when the mass is subsequently cooled. If the right amount of fondant is added at the right temperature, a sufficient number of seeds remain to make fudge with the right texture.

If too many seeds dissolve away, the remaining crystals grow larger and grainy fudge is produced. Thus, the temperature of addition is critical to getting a good product.

The amount of crystals, or crystalline phase volume, is also important to texture—the more crystals, the harder and more crumbly the fudge. Water content and the ratio of sucrose to corn syrup are the two factors that most affect graining. In fact, the higher ratio of sucrose is what typically differentiates fudge from caramel.

What makes fudge get hard? Water—or rather, too little of it. The sugar syrup for making fudge is cooked to the firm ball state (see Chap. 12), or about 244 °F, to leave 10–12 percent moisture. If cook temperature goes a little higher, more water is evaporated and more sugar crystallizes out upon cooling. The result is fudge that's as hard as the proverbial biscuit.

Fudge can also dry out if left open to air. Moisture loss to the air causes a perfectly soft fudge to become as hard as a rock.

Fudge is also one of the few confections that can support mold growth, at least under certain circumstances. Typically, fudge is resistant to mold growth because the water is tied up with the sugars and not available for the microorganisms to use to grow. When fudge gets moldy it's often related to moisture condensing on the surface in humid conditions. The water dissolves some sugars from the fudge, making a nutrient-rich broth ripe for mold growth. To prevent surface mold growth, confectioners cover the fudge carefully and completely so no moisture condensation occurs.

Just like uses for the word, fudge comes in a wide variety depending on texture and flavor. Whether you're an artisan fudge-maker out to make a special treat or an engineer calculating

34

English Toffee

A while ago, a small candy maker visited us to talk about shelf life of her English toffee. As a small business, she makes individual batches of toffee in her kitchen (state health approved, of course) and sells them in various regional outlets as well as on-line. Recently, she has begun to expand into larger distribution centers.

Based on experience, she knows that her candy is still good even four months after making it. After that, though, she can't guarantee its quality. The larger outlets are asking her for a six month shelf life in order to put them in their national outlets. She was asking how she might do that.

What is English toffee anyway? Her label says the ingredients are milk chocolate (the coating on top of the toffee), sugar, butter, cream, salt, and nuts. Essentially, American English (is that an oxymoron?) toffee is sugar, dairy, and nuts cooked to a high temperature, and sometimes coated or layered with chocolate. It differs slightly from peanut brittle, primarily in the fat content—English toffee is substantially higher fat content.

When we make English toffee in our lab, we essentially just mix butter and sugar, cook it up to 260 °F, add unroasted almonds and cook to 305–310 °F until it's nice and browned. Cooking to a high temperature drives off the appropriate water, up to the hard crack stage (see Chap. 8), promotes both types of browning reactions for color, flavor and aroma (see Chap. 28), and allows the nuts to roast in the cook as well. The cooked mass is then poured out onto a cold table, spread out and allowed to cool in a thin layer. As it cools, it sets into a glassy sugar matrix with fat globules and nuts dispersed throughout.

R.W. Hartel and AK. Hartel, *Candy Bites*, DOI 10.1007/978-1-4614-9383-9_34,   135
© Springer Science+Business Media New York 2014

Some toffee recipes call for a little sodium bicarbonate (baking soda) to be added right at the end of the cook. This sets off an acid–base reaction, generating small carbon dioxide bubbles, which are also dispersed throughout the sugar glass after cooling. The aeration lightens the texture a little, making the toffee easier to bite through.

Numerous variations of English toffee exist. One version in England, not called English toffee, is Bonfire toffee. It's essentially the same as English toffee but flavored with molasses. Another version sold commercially in the United States is Almond Roca buttercrunch toffee, made by Brown & Haley of Seattle, WA. One claim to fame of Almond Roca is that it's sold in a pink tin.

In fact, that tin serves two purposes, as do most food packages. The tin serves to market the product and to preserve it at the same time. By sealing the candy in air tight tins, the English toffee lasts far longer than if it wasn't in the tin.

What causes English toffee to go bad? One of two things. For one, if you store English toffee in humid air, it will quickly pick up moisture and get sticky. Eventually, it will be too sticky to eat. The tin can helps protect against moisture uptake. The second mechanism for end of shelf life is that the fat goes rancid. Lipid oxidation leads to stinky off-flavors as the fat is broken down. It's driven by the presence of oxygen, so minimizing exposure to air by storing in air-tight tins means lipid oxidation doesn't happen and shelf life goes up.

Because of the extended shelf life, tins of Almond Roca toffee were sent to soldiers during World War 2. Their web site claims it was even taken along on Mt. Everest expeditions by Sir Edmund Hillary (although I doubt he schlepped a tin can up to the summit). As long as the tins remain closed and sealed, the candy lasts a very long time.

What tools would our local English toffee maker have at her disposal to extend her shelf life out to six months? Both formulation and packaging options exist. Let's look at potential formulation changes first.

English toffee is a sugar glass and is notoriously hygroscopic, it quickly sucks water out of the air. Unfortunately, there's not much leeway for replacing sugar without completely changing the product. One option is to use a sugar alcohol like isomalt or maltitol, which are slightly less hygroscopic than sugar. They cook up much like sugar, and have a similar brittle texture, and so may be a reasonable alternative in that sense. But, they don't undergo browning reactions, so the toffee color, flavor and aroma have to be added separately. So, other than as a sugar-free alternative, these sweeteners won't cut it. A packaging solution is the best approach to slow moisture sorption and stickiness.

But lipid oxidation is a different story. It turns out that certain types of fat are more prone to lipid oxidation than others—typically, more unsaturated fats go rancid faster than saturated fats. Milk fat has a fair degree of saturation, but there's way too much unsaturated fat available and, if exposed to oxygen, will go rancid fairly quick. We figure the English toffee we make in the lab, wrapped only in plastic sandwich bags, goes rancid within a few weeks (fortunately, it rarely lasts that long).

Replacing the butter with a more saturated fat is an option. Vegetable fats with higher melting point are more resistant to lipid oxidation. But they have a down side as well, primarily in the loss of the cooked butter flavor notes. Additional flavors can be added to the batch, but that's a lot less natural. Vegetable fats also don't look as appealing on a label as butter.

So, it appears that formulation changes don't provide acceptable options to extend shelf life; let's look at packaging. Essentially, anything that provides a barrier to both water vapor and oxygen would help extend shelf life. Wrapping individual candy pieces in a foil wrapper is a good start. Then packaging those foil-wrapped pieces in an outer bag as another barrier helps more. And then sealing the bag into an air-tight tin will really protect the candy from the environment. Although that's a huge expense, it's a guaranteed approach that's been proven to work.

Another trick is to coat the toffee in chocolate. As a fat-based composite material (sugar, particles, cocoa solids, milk powder, etc.), chocolate is actually a decent water and oxygen barrier, besides being a delicious complement to English toffee. It's no wonder that many toffee products are associated with chocolate in some form or another. However, chocolate is not a perfect barrier to either water or oxygen and the toffee would eventually go bad.

Further, to solve our candy maker's problem using this chocolate approach would require her to change the nature of her product. Instead of having a chocolate topping (with nuts), she would have to completely encase her toffee in chocolate. That's also not desirable.

As is often the case when a small company wants to grow larger, there is a business decision to make. She can either stay out of the bigger markets with the six month shelf life requirement or change her product or package in some way to fit their needs.

35

Gummies and Jellies

Throughout history, our ancestors have sought out compounds that can thicken liquids. In cooking, we thicken broth to make gravy, thicken fruit juice to make Jell-o, and thicken juices into aspic. The medical field also thickens water and other fluid drinks to assist patients who suffer from dysphagia, or swallowing problems.

Where do all those thickeners come from and which ones can be used to make gummy and jelly candies?

To some candy purists, there is a clear distinction between a gummy and jelly candy. Gummy candy, as the name suggests, has a gummy texture. Not quite like chewing rubber bands or calamari, but certainly more elastic than any other soft candy. Since gelatin is the only material that gives that texture, by definition, gummies are made with gelatin.

A jelly candy then is one made with anything other than gelatin—pectin, starch, agar, gum acacia and so on. Each has a different texture, but none are elastic like gelatin. In fact, it's the "holy grail" of the soft candy maker to find something with gelatin-like texture without actually using gelatin. The animal origin of gelatin (see Chap. 39) limits its use in certain diets, but its unique texture cannot be replaced by other gelling agents, at least so far.

The primary characteristic of gummy and jelly candies is the use of a stabilizer, or gelling agent. These large protein, gum or polysaccharide molecules interact in a sugar solution to form a 3-dimensional network that holds in the fluid sugar solution. At 18–20 percent moisture, the sugar solution held within the gel network would still be sufficiently fluid to flow if not constrained somehow.

R.W. Hartel and AK. Hartel, *Candy Bites*, DOI 10.1007/978-1-4614-9383-9_35,
© Springer Science+Business Media New York 2014

Probably the most common jelly candies are made with starch, usually from the corn plant. Generic products made with starch include orange slices, peppermint leaves, jelly rings, gum drops and Turkish delight. Some branded candies made with starch include Dots, Jujyfruits, Swedish Fish and Chuckles. Starch jellies have a chewy, but not elastic, texture that can be anywhere from soft to hard, depending mostly on water content but also on the amount and type of starch used (see Chap. 37). Because of the way starch molecules interact to form a gel, starch jelly candies are translucent to opaque. You can't see through them.

Another popular jelly candy stabilizer is pectin, a biopolymer rich in galacturonic acid that makes up a part of cell walls in plants. An extract from apple or orange skins, pectin provides natural thickening power for jam and jelly candies. When used in a jelly candy, pectin produces a unique texture. Although some consider a pectin jelly to be slimy, it's characterized by a short, tender, even brittle, texture. Also, since the junction zones of the pectin gel network are small, they don't scatter light and a pectin jelly candy is amazingly clear. When we pour a pectin jelly onto our cold table in the lab, you can see every detail and scratch of the table right through the gel. As the instructor in our candy school says, if you poured it onto a newspaper, you'd be able to read right through the candy. Turning pages would be tough and it would make a mess, but the image gets the point across.

Pectin actually comes in various forms, with different chemical make-up and setting properties. For jelly candies, what's known as high methoxyl pectin is used because its properties lend themselves to sweets. To get high methoxyl pectin to set, high sugar content and acid are needed. Sugar molecules, through their interaction with water, enhance interaction among pectin chains, while acid neutralizes the charged hydoxylate groups and promotes gelation. When a candy maker makes a pectin jelly, the last ingredient added to the batch is usually citric acid to reduce the pH. This causes the pectin to set up within minutes so it's imperative that the time from acid addition to mold filling be very short. If not, the candy in the lines solidifies and someone loses their job. And someone else has the unenviable task of clearing out the blocked lines.

Another plant-based stabilizer found in some jelly candies is agar, sometimes redundantly called agar agar. Agar comes from the cell walls of red seaweed, so it is also a natural stabilizer. More often used as growth medium for colonies of microbes, agar finds specialized use in soft jelly candies. Some fruit slices may be found made from agar, but they're hard to find. With a texture reminiscent of pectin, agar jellies are also soft and tender.

Gum arabic, also called gum acacia, is also sometimes used as a thickener or stabilizer, one that finds specialized use in confections. Derived from the sap of the acacia tree, gum arabic is also sold as a dietary supplement for reducing cholesterol and for promoting weight loss. It's also used in emulsified soft drinks, like Mountain Dew, to prevent the flavor emulsion from breaking. In candy products, the main use of gum acacia was Pine Bros cough drops, known as the "softish" throat drops due to the unique texture imparted by gum acacia. Unlike other stabilizers, gum acacia provides a firm texture that's nearly impossible to bite through, hence the softish texture. Pine Bros throat drops have recently made a comeback from the dead candy list.

Although there are numerous stabilizers available to provide different and unique textures to soft candies, candy makers are always looking for something unique. Blending the different stabilizers allows the candy maker to produce soft candies with textures that vary from the elastic texture of gelatin to the tender bite of pectin. Adding starch or pectin to gelatin reduces the elasticity of the gummy candy. Although numerous mixed-stabilizer candy products are on the market, these are most often found in other candy-related products—fruit snacks and chewy vitamins.

What do we call these mixed-stabilizer candies? Anything the marketer wants. As with many things, the definition of something is often blurred through common use, where some marketing person thinks they can get a leg up on the competition. This is certainly the case with gummy and jelly candies.

36

The Starch Mogul

How many definitions are there for "mogul?" First, there's the bump or small hill on a ski slope that provides a challenge for the expert skier and a nightmare for the novice. A mogul is also a very important person—a high muckety muck, the big cheese, the head honcho, and so on. A mogul is also anyone related to the Mughal (or Mogul) empire. But say mogul to any candy maker and they'll ask whether you mean the processing equipment or the company CEO, a business mogul.

In the previous chapter, we learned what gummy and jelly candies are made of. Here, we'll delve into exactly how they're made. And that brings us to the starch mogul or the system for depositing candies into starch to create forms. I had to ask around to find out how the starch mogul got its name. Fortunately, I thought to ask Jim Greenberg, co-President of Union Confectionery Machinery in Brooklyn, NY.

Jim says the starch mogul was first developed in the 1890s by National Equipment Company. The system was developed, named and patented by an engineer at National Equipment. Although the patent has long expired, it started something that continues to grow to this day. The mogul provided a real advance at the time, from hand-made goods to continuous production on a large scale.

Whether making gummy or jelly candies, the process generally follows the same protocol. First, the syrup is cooked to the appropriate temperature to get the right sugar content, or °Brix (read as degrees Brix), a unit based on refractive index of a sugar solution. Typically, we're looking for 78–80°Brix, which corresponds to approximately 18–22 percent water content (since refractive index

R.W. Hartel and AK. Hartel, *Candy Bites*, DOI 10.1007/978-1-4614-9383-9_36, © Springer Science+Business Media New York 2014

depends on the type of sugars present). At this point, the sugar syrup is sufficiently fluid that it can be deposited into a form to take on the candy shape. But at this water content, the candy would be really soft when the stabilizer sets up into a gel and that's why starch depositing is used—the starch helps to dry out the candy further.

A tray or board is filled with dried starch powder treated with a little oil. The oil allows the starch granules to hold the shape when a depression is made in the starch, in much the same way as a little water in beach sand allows the sand to hold a shape. A depression is formed in the dried starch by pressing a shape, or press board, firmly down into the flat layer. The shape can be almost anything from a bear's body to an orange slice to a hand (or face) print. Seems like every year in candy school we get someone willing to get powdered starch on their hand to make a gummy hand. It's rare that someone sacrifices her face to make a gummy mask.

After the fluid candy syrup is deposited into the starch mold, it's put into a curing room to allow the stabilizer to set. The starch also pulls some moisture out of the candy to help with gelation. For starch-based jelly candies, the curing room is warm, about 110–110 °F, whereas gelatin based gummy candies have to be cured at cooler temperatures because of the low melting point (see Chap. 39). After curing, the candies are removed from the starch, blown clean of excess starch with forced air, and then either sugar sanded or oiled.

While the entire process used to be done by hand, the automated starch moguls were developed to accomplish each step much more efficiently. As noted above, the continuous starch mogul was developed in the 1890s, during the era when everything was being automated, from cars to sweets. The entire gummy or jelly candy process could then be accomplished in one huge machine.

The starch mogul consists of several distinct sections with a conveyor continuously moving starch trays from one end to the other and each step of the process taking place at a different stage. On one side of the conveyor is the feeder section. Here, candy-filled starch trays from the curing room (candy that was made the day before) enter into the mogul. With modern automation, up to 35 trays per minute can enter into the system.

The primary working area of the mogul is called the starch buck. Here, the candy-filled trays are turned upside down over a screen to sift out the candies. The candies go one way for further processing (sanding or oiling) while the starch goes another way to be dried prior to reuse in the mogul. The trays are turned again and filled with dried powdered starch, recycled from the dryer. The starch surface is scraped clear to provide a clean, virgin surface for jelly artistry. A huge pressboard containing rows and rows of the imprint shape is pressed into the clean white starch surface to create the molds.

Although most of the starch is collected, dried and reused, some of it ends up on the floor and everywhere else around the plant. A starch mogul is a messy place, with starch blowing all over everything. It's no wonder that a significant amount of fresh starch has to be added periodically to keep up with that loss.

After leaving the starch buck, the molds are ready for filling. In the lab, we use small hand-held funnels to fill the candy into the starch molds. In a mogul, multi-head depositing nozzles fill row upon row of candy shapes. Multiple colors can be deposited at the same time to simplify making mixed color products (think of all the different colors of gummy bears in one bag). Accurate weight systems control the exact amount of candy deposited in each shot and special suck-back nozzles prevent syrup tails from dragging between deposits. To speed the process, the hoppers and depositing nozzles swing with the trays as they're indexed on the conveyor.

Once the trays are filled with fluid candy, the last stage is unloading the trays from the mogul. Trays filled with the still-wet candy syrup are stacked atop each other onto racks that automatically transport the candy to the curing room, where the candy is allowed to set overnight. The cured candy trays are pulled out the next morning and fed into the mogul, starting the process again. In the largest candy companies, the starch mogul runs for days on end, producing tons of candy each hour.

Development of automated starch moguls had a huge impact on candy production. What had traditionally been done batch-wise and by hand could now be mass produced by the bazillion. In fact, like fast food meals, the mogul has been super-sized to the point where Jumbo moguls can produce on the order of 20,000 pounds of candy per hour. That's a lot of gummy bears.

37

Swedish Fish and Starch Jelly Candies

What makes Swedish Fish different from Turkish Delight, Jujyfruits different from orange slices, or spice drops different from Dots? Although they're all made from starch, the texture of these candies can vary widely.

Swedish Fish, a relatively firm jelly candy, was brought to the United States in the 1960s by a Swedish firm looking to expand their market. They must have figured that Americans thought of fish when they thought of Sweden. A few years ago we had a visiting student from Sweden. She wasn't able to shed light on the name, although took a liking to the American version while here.

Turkish Delight, a much more delicate starch-based jelly candy, has been around for centuries. It may in fact be one of the oldest of confections, supposedly developed 500 years ago for an Ottoman Sultan. It traveled to Britain in the eighteenth century where it became a treat for high-class society. It still enjoys great popularity in certain regions.

The texture of these candies is partially attributed to the nature of the starch used to make the confections, although the final water content also plays an important role (see Chap. 38). Here we'll focus on starch and its properties for making jelly candies.

To make starch-based jellies, we need sugar, corn syrup and starch, along with some colors and flavors. Fortunately, starch is ubiquitous in nature. It's found in a wide range of plants as an energy storage medium. One outcome of photosynthesis is the generation of glucose from carbon dioxide. The plant then polymerizes glucose into starch molecules, which are then packed

R.W. Hartel and AK. Hartel, *Candy Bites*, DOI 10.1007/978-1-4614-9383-9_37,   147
© Springer Science+Business Media New York 2014

tightly into a granule and stored for future needs. Those starch granules, then in turn, provide energy for us humans when we eat those plants, with our body essentially reversing the process to liberate energy for our bodily functions.

Starch-producing plants include corn, potato, tapioca, wheat, sorghum, rice, and a host of others. Through processing, we separate the starch granule from the plant and dry it into a starch powder. Corn starch powder, for example, is a fine-grained, white powder that can be used in a wide variety of applications, including making Swedish Fish and Turkish Delight. In fact, it's used twice to make jelly candies (see Chap. 36).

The starch granule is an interesting product of nature. The plant has developed this highly sophisticated method of storing starch by producing a complex structure, the starch granule. Each starch granule is a mixture of the two types of starch—the straight-chained amylose and the branch-chained amylopectin. Stacking amylose molecules would be like stacking 2×4's; they form a compact pile with the boards (amylose molecules) in close contact. Amylopectin, on the other hand, stacks more like numerous loose branches from pruning a tree. Because of the different angles that the branches shoot from the main trunk, they don't pack very well at all. You have to jump on the pile to break the branches to get them to compact.

These arrangements are important because both amylose and amylopectin are packed together within the starch granule, at different ratios depending on the plant source. The most efficient organization is layers of the branched amylopectin interspersed with the straight chains of amylose in a semi-crystalline organization. The birefringence, or Maltese cross, when looking at starch granules under a microscope with polarized light is classic evidence of this type of semi-crystalline arrangement.

The starch granule is essentially impervious to cold water—the tight-knit, semi-crystalline nature of the starch molecules prevents water from penetrating into the granule and allows the plant to store energy in a compact form. The nature of the starch granule, small and irregular in shape, leads to some interesting properties.

Mix about 80 percent dried corn starch granules with about 20 percent water and see what you get. This mixture, sometimes called ooblek or mind pudding, has unique properties. It acts solid-like when you run across a pool of it (like on the Ellen Degeneres Show), but it becomes fluid as soon as you stand still—you sink. It's known as a shear-thickening fluid, when you move it rapidly it becomes solid (like rolling a ball in your hands), but when you stop moving or stirring, it becomes liquid again (the ball flows in your hand when you stop rolling).

You'd think someone would have turned this shear-thickening behavior to good use, but so far no one has been able to take advantage of this unique property to make a commercial candy product. How about making a version of Nickelodeon slime into a candy that had these unique properties?

Starch granules change quickly once the water is heated, however. Water can now penetrate into the granule, causing interesting changes that are harnessed to make jelly candies. When a sugar syrup containing anywhere from 9 to 15 percent starch granules is cooked to boiling temperatures, numerous changes take place that allow us to make a wide range of candy products.

First, as the system heats up, the crystalline regions of the granule melt and the water molecules now penetrate into the tight structure. The granule swells as this happens to allow room for the water to move between individual starch molecules. This swelling causes the viscosity to go up and is responsible for thickening of gravy. Eventually, the straight-chained amylose molecules are able to diffuse out, departing the swollen granule for the broader expanse of the sugar solution. Eventually, the entire granule disappears, leaving a soup of starch molecules in solution with the sugars. This process is called pasting of the granule.

The resulting mixture of sugars, starch molecules and water is the basis for making starch jelly candies. This mixture is what's poured into molds to form candy shapes, from orange slices to fishies. When cooled, the starch molecules, primarily amylose because of its structure, turn into a gel to provide the firm texture of a jelly candy. This part of the process is called gelatinization.

As always, humans look for better ways to do things. It's no different in making jelly candies. One problem with cooking native corn starch is that the hot fluid syrup is too viscous to fill easily into molds to make candy shapes. We could raise the temperature a little more or leave more water in the syrup, but these cause other problems. One solution is to modify the starch a little to make it less viscous after the granules were cooked out. By breaking down some of the longer starch molecules into shorter segments, using either acid or enzyme treatment on the native starch granules, the viscosity of the cooked slurry is reduced, making it much easier to work with. This "thin-boiling" starch provides a distinct processing advantage.

And sometimes the texture of the gel structure isn't exactly what's desired in a candy. Jujyfruits require a different texture than orange slices or Dots. We can distinguish these different textures by either using more starch in the candy or by changing the type of starch. More starch in the mixture means more amylose molecules to come together to form a gel. More interaction points means a harder gel. Because tapioca starch has lower amylose content than normal (dent) corn starch, it forms a softer gel. We can also use high amylose starch, specially bred to have up to 70 percent amylose, to generate a rigid gel and a firmer candy.

So although there must be hundreds of different starch jelly candies on the market, through control of the starch, we can tailor the texture to whatever we desire in our products. From the hard and tough jujube to the soft bite of a fresh Dot to the chewy Swedish Fish, we can give the customer whatever she wants through control of starch chemistry.

38

Dots and Orange Slices

"Sweet, I found a box of Dots back here in the drawer, want some?" says Joe. Claire replies, "Wait, how long's it been in there?" "I dunno, probably a few months," shrugs Joe. Claire thinks for a moment, back to her class on candy science, and then says "You can have them, I'll go get some new ones."

What Claire knew was that Dots in a box have a relatively short shelf life, well at least if you want to keep all your teeth. If you don't mind Dots that chew like Jujyfruits, then well-aged Dots are just for you.

Both Dots and Jujyfruits, and Jujubes for that matter, are starch jelly candies. They're essentially thick and gooey, colored and flavored sugar syrups entrapped in a network made from starch that's been gelatinized. The texture depends on a lot of things, including the type and amount starch used in the confection (see Chap. 37), but water content also plays an important role in the texture of starch candies, indeed for all types of gummy and jelly candies.

When we make Dot-like candies or orange slices in candy school, it takes a couple days to complete the process. The first day is for cooking the starch slurry to the right water content and depositing the hot liquid candy into the starch molds (see Chap. 36). These are put into a warm curing room to set overnight. The corn starch that makes the molds also dries the candy out a little. The next day, we remove the solidified candy from the starch powder, blow off any remaining starch granules, wet the surface with live steam, and throw the candy into a tumbling pan of

R.W. Hartel and AK. Hartel, *Candy Bites*, DOI 10.1007/978-1-4614-9383-9_38,
© Springer Science+Business Media New York 2014

sanding sugar. The result is one of the most delicious orange slices you've ever had.

Why are they so good? Partly because we add extra flavor and acid, but mostly because they're really soft and tender when they're just made. They're so soft they almost melt in your mouth, quickly releasing the delicious orange flavor. Even people who don't particularly enjoy orange slices think they're delicious. Unless we get it wrong.

Sometimes, the candies get left in the starch trays for an extra day or two. Because they've been losing water to the surrounding starch powder for a longer time, they've dried out more and are much firmer than the orange slices only left to cure overnight. We've learned that water content tracks directly with firmness of our orange slices. Or Dots or Black Crows, or any jelly or gummy candy for that matter.

Dots were originally produced in 1945 by the Mason Candy company, who also produced Mason's Black Crows, the licorice-flavored version of Dots, as early as the late 1800s. Why Black Crows preceded the fruit-flavored Dots is unclear but probably related to the preference for licorice flavoring years ago (see Chap. 40). Both Dots and Black Crows are currently made by Tootsie Roll Industries.

The candy manufacturer has control over the initial water content, through the sugar syrup poured into the molds and the time spent in the curing room. But once the candy is packaged and on the trucks for distribution, it's out of their hands. The water in the candy is still moving around during storage and distribution, usually with undesirable results.

Why do Dots get harder over time, making the old box that Joe found in the drawer almost inedible? It's thermodynamics. The water molecules within the Dots want to equilibrate with water molecules in the air around it, often leading to the end of shelf life.

If Dots were stored in an environment where the water in the air had the same "activity" as the water molecules within the Dots, there would be no net exchange of water molecules, they'd be at equilibrium. Equilibrium doesn't mean that there's no exchange of water molecules between air and Dot, just that the same number of molecules go one way as the other—no net change. Candy scientists would say that the Dots were being stored at their "equilibrium relative humidity." For soft jelly candies like Dots and orange slices, that relative humidity (RH) would be about 50–60 percent.

Technically, if the Dots were stored in air with higher RH, there would be a net migration of water molecules from the air into the Dot, most likely resulting in a sticky surface. This rarely happens in the temperate climates of North America, where average RH over the year is lower than 50 percent. Sure, there are some summer days when you start sweating immediately after taking a shower because it's so muggy (high humidity), but on average over the year, the atmospheric RH is lower than the equilibrium value of a Dot. That means there is a net loss of water from the Dot due to the thermo-dynamic drive of the water molecules to equilibrate with the air.

The result is hardened Dots, as Claire knew, and as Joe will find out when he tucks into that old box of Dots he just found. The dried Dots will be a lot firmer and harder to bite through than fresh ones from the store.

What options do candy makers have to stop or at least slow down the moisture loss that leads to the end of shelf life of their products? The first line of defense is the package. If Dots were sealed in foil wrapping package, they'd last a lot longer. Candies like Pop Rocks and cotton candy are packaged like that because they pick up moisture so fast from the surrounding air. But not Dots, or any gummy and jelly candy for that matter. They're packaged in some sort of plastic film or maybe in a cardboard box wrapped with plastic film. Cardboard and the thin plastic films used for most candies are minimal water barriers at best. They don't provide much of a barrier to the water molecules moving around trying to find their equilibrium.

Another approach for some jelly candies is to apply oil and wax to the surface as a water barrier. Look at the ingredient list of products like Jujyfruits and Swedish fish and you see mineral oil and carnauba wax near the end. These form a thin coating of oil/wax, which are hydrophobic (don't like water) materials. This

39

Gummy Jigglers

Do you know what holds the fire-starting compounds on the business end of a match? It's the same material we use to make gummy bears (or gummi bears, depending on your inclination, both are acceptable).

Gelatin, a protein derived by breaking down collagen, appears regularly in our daily lives. It's used as a biodegradable glue for things like phone book bindings and sealing corrugated cardboard. Softgel capsules made with gelatin protect vitamins and other pharmaceuticals until released when needed (in our stomach) in the same way that the gelatin skin on a paint ball protects dollops of paint until it splats on a target. It's used for ballistic testing—shooting bullets into a gelatin brain is better and more informative than using cadavers. It also provides a source of protein and amino acids in various creams and cosmetics. Gelatin is also used to clarify wine, beer, and juices. Gelatin microcapsules hold ink for carbon-less copy paper. Old-time photographic films used gelatin to hold silver halide crystals on a film.

Gelatin also is widely used in foods. It provides the jiggly character of Jell-O and aspic (a savory form of Jell-O). Gelatin gives the unique springy texture to marshmallow. It's still some-times used as a stabilizer in ice cream (like the Babcock Hall Dairy plant at the University of Wisconsin-Madison). And last but not least, it provides the elastic chewy character of gummy bears, worms, tarantulas, hearts, and all sorts of other gummy candies.

Traditionally derived from hides and bones of pigs and cows, gelatin is a breakdown product of the structural protein collagen. You might recognize collagen as the gristle in a cheap cut of meat.

If you slow cook that meat all day in a crock pot, it gets more tender because the collagen is broken down into smaller molecules, somewhat like gelatin. What gives gelatin its unique elastic characteristics is the capability of the molecule to form strong junction zones as it gels, trapping fluid within the network of cells formed by the junctions.

According to all accounts, the first use of gelatin in gummy confections is the gummy bear, originally developed in 1922 in Germany. A candy maker from Bonn named Hans Riegel is said to have made "dancing bears" out of gelatin. He named his company Haribo, from the first two letters of his name and home city (Ha-Ri-Bo). We now know Haribo as one of the primary producers of gummy candy products, including the original gummy bears, but there are numerous other versions available these days.

Not all gummy bears are created equal. Some are more elastic than others, some harder and others softer. In general, Europeans have a taste for really elastic gummy candies while Americans like their gummies less gummy. Compare Haribo with an American brand like Black Forest (must be the Black Forest region of Chicago, where they were first made?). Although the American version is still quite gummy, it's nowhere near as elastic as the European version.

What causes the differences in these products are a combination of things, but much of it comes down to the gelatin characteristics. Not surprisingly, gelatin comes in numerous varieties, each with its own characteristics.

Gelatin makers and users specify gelatin according to its "bloom" strength. To measure bloom strength, gelatin scientists carefully prepare samples and measure how much force is required for a probe to penetrate a specific amount. The protocol is standardized so bloom strength is a universally accepted number. A bloom strength of 250 means that it took 250 grams of applied force to penetrate the specified distance under the set conditions. Typical bloom strength of gelatin for gummy candies varies from 200 to 250 with gelatin levels of 7 to 9 percent.

A gummy bear made with 9 percent of 250 bloom gelatin will be significantly more elastic than one made with 7 percent of 200 bloom strength. In comparison, gelatin desserts are made with similar bloom gelatin but at significantly lower concentrations—enough to make a jiggle but not enough to prevent a spoon from slicing through.

A unique characteristic of gelatin is that it makes a thermoreversible gel. It has a specific melting point and when heated above that temperature, the junction zones unfold and the gel structure dissipates into a viscous fluid. Cool the gel back below its melting point, the gelatin molecules once again form junction zones and it re-solidifies back into a gel.

If you heat a gummy bear above its melting point while it's sitting on a flat surface, it'll melt and flow out into a puddle (liquid finding its level). Once cooled back down again, all you're left with is a gummy pancake—no sign of a bear any longer.

The melting point for gelatin candies varies between about 35 and 40 °C (95 and 105 °F), depending on conditions. That's great because it (mostly) melts in your mouth, but not so great for stability during distribution. Shipping a bag of gummy bears in the summer without cooling isn't a good idea. You're likely to receive a bag of package-shaped gummy sweetness, perhaps without even any memory that bears were present.

The growing demand for gelatin—it's projected to grow by up to 6.75 percent through 2018—means that there will be more competition for gelatin. Although food is the major user of gelatin (nearly 30 percent of the total market in 2011), other users are likely to grow considerably, especially the nutraceutical and pharmaceutical products. While the health benefits of eating straight gelatin (remember the Knox gelatin ads?) have been sort of forgotten over the past few decades, the idea that gelatin is good for you is making somewhat of a comeback. Improved digestion, less creaky joints, ageless skin, improved sleep and, of course, better finger and toenails are all reasons for taking gelatin before bed.

Because of the growing demand, and to some extent because it comes from an animal source, scientists have been searching for alternatives to gelatin for decades. Recent work has been successful at genetically engineering a source of gelatin, but whether or not people would eat gummy bears made from recombinant gelatin is not so clear. Fish gelatin is available, but doesn't pack quite the same jiggle.

For the most part, research into elastic gelatin-replacers for gummy candy has been disappointing at best, although hydrocolloid chemists continue to seek one of the holy grails of the candy industry—a kosher material with gelatin's jiggle.

40

Black Chuckles

Black Chuckles—a cousin of a devious laugh, a subdued muahaha. No, we're just talking about the black piece in the center of the 5-pack of jelly candy known as Chuckles. You know, the candy that "even the name says fun." You may think these are retro candies that aren't made any more, but you'd be wrong. Chuckles are still sliding off the conveyor at the candy factory, laughing all the way to the bank. Well, they're at least making enough money to warrant their continued production nearly 100 years after their start.

They were originally developed in the early 1920s by the Amend Company in Chicago, IL, but have been bought and sold a lot in past years. After being acquired by Nabisco and then Hershey, the brand was sold to Farley's & Sather's, which recently merged with Ferrara Pan Candy to become Ferrara Candy Co. Chuckles are a good example of how a candy can be bought and sold over the years, yet still have a continued presence (see Chap. 4).

What's your favorite flavor of Chuckles? In one pack, you get five candies with five different flavors, cherry, lemon, licorice, orange and lime. You might be surprised that the black one, licorice, is the favorite for many people. That's right, there are people that actually like the black licorice Chuckles.

In fact, out of the 11 college students polled in our candy science class one year, two said the black Chuckle was their favorite (or at least co-favorite). On the other hand, five said they absolutely detested the black one, although that included one who said she hated them all. The remaining three students were noncommittal

R.W. Hartel and AK. Hartel, *Candy Bites*, DOI 10.1007/978-1-4614-9383-9_40,
© Springer Science+Business Media New York 2014

about the black one. Clearly, although the black one has supporters, more people dislike them, and dislike them intensely.

There was a time when licorice-flavored jelly candies were a lot more popular. When I was a kid, my father used to buy entire bags of black, licorice jelly beans. A whole bag of just black jelly beans! You can still buy them, although it's difficult to find them in the store any more. Tastes have changed.

To a kid with a sweet tooth the size of an elephant's tusk (yes, it's really a tooth, and the largest one at that), even black jelly beans were a treat. I'd eat anything sweet, even sugar cubes (see Chap. 19). When I needed a fix, which was almost every night, I'd creep downstairs in the middle of the night to steal a handful of those black jelly beans from my fathers stash. When morning came, I'd then spend the entire day sweating that he'd notice the bag was lighter and look at me as the most likely culprit. Fortunately my handfuls weren't that big. Either that or he knew and let it slide. Nah, if he knew I was ripping candy, he'd have had the holzloeffel (wooden spoon), his preferred means of beating some sense into me, out in record time.

Black jelly beans, Chuckles (and their cousin, jelly rings), and licorice Crows are examples of starch jelly candies flavored with licorice. Other starch jelly candies, not licorice-flavored, include gum drops, spice drops, and orange slices. Dots, Jujyfruits and jujubes are also starch jellies but with a harder texture because they have lower water content, especially when they get older. But that's another story (see Chap. 38).

Starch jellies are made with sugar, corn syrup, and starch. Of course they also contain colors and flavors, like black licorice. The starch gel is what makes these candies unique.

Although a Chuckle looks like a solid piece of candy, it's really a liquid sugar syrup with up to 20 percent water held in place by the starch gel. If the starch wasn't there to hold in the liquid sugar, the syrup would just flow out across the table, making a sticky Chuckle puddle.

Look carefully at a handful of corn starch powder. Each starch granule is only about 10 μm in size, but it's packed full of starch molecules. With water and heat, those starch molecules can be coaxed out of the granule into solution. In a fully cooked starch slurry for making black Chuckles, all of the granules are completely disintegrated, with the individual starch molecules swimming in an ocean of sugar water. While this sweet mixture is still hot and liquid, it's poured into molds with the distinctive tire tread pattern typical of Chuckles. The top of the mold is open so the liquid candy mass flattens as it seeks its own level (the definition of a liquid). Look carefully at a Chuckle—it's easy to see which side is the top and bottom of the mold.

The liquid candy in the mold cools and the starch molecules solidify, somewhat like over-gelled gravy. The process of starch molecules forming a gel is called gelatinization. The interaction of the long starch molecules produces a gel-like network that entraps the fluid sugar solution within a 3-dimensional structure.

In the original Chuckles plant back in 1921, the sugar/starch granule slurry was cooked in batch kettles over open flames. They were then deposited by hand, allowed to cure, and then sugar-sanded, again all by hand. Modern jelly candy cookers continuously cook the starch slurry and can make a bazillion more Chuckles in a day than the original batch process. Sometimes called jet cookers because of the noise they make, modern cookers inject hot steam into the sugar/starch slurry. The hot steam condenses directly into the slurry, causing the starch granules to quickly hydrate, expand and then disintegrate. Almost nothing is left of the original starch granule, with almost all the starch molecules out in the syrup.

Back in 1949, they even made an all black licorice flavor pack of Chuckles. As the advertisement read, "5 luscious slices for five cents." Because tastes have changed over the years, the black Chuckle packs are no longer made. But some people still favor the black one, and they have the last laugh when they get to eat yours. Are you going to eat that?

41

Fruit Snacks

Are fruit snacks candy? Oh, there's a loaded question. Some of the companies that make fruit snacks don't consider themselves candy companies, primarily because they make a wide range of different products (from cereal to granola bars). However, some companies that make fruit snacks are also clearly candy makers, with other products that fall squarely into the candy aisle in their portfolio.

Let's look at the labels of two brands of fruit snacks and see how they compare. One brand, from a company that wouldn't consider itself a candy company, contains: juice from concentrate, corn syrup, sugar, modified corn starch, fruit pectin, citric acid, dextrose, sodium citrate, malic acid, color, sunflower oil, Vitamin C, natural flavor, carnauba wax. Another, from a company who produces other candies, contains: juice from concentrate, corn syrup, sugar, modified corn starch, fruit puree, gelatin, citric acid, lactic acid, natural & artificial flavors, Vitamin C, Vitamin E, Vitamin A, sodium citrate, coconut oil, carnauba wax, colors.

Both lead off with fruit juice from concentrates rather than corn syrup and sugar, although those are there as well. Fruit juice concentrate sounds much healthier than sugar and corn syrup, right? But let's look a little closer. The commercial products that these companies most likely use are essentially clarified and deodorized juice concentrates. For example, the juice from either apples or pears (or both) is collected, clarified and concentrated, which generally also removes the aromas. That's actually a good thing because they'll put back in flavors to suit. That is, rather than using raspberry juice concentrate to make raspberry-flavor fruit snacks, they'd use deodorized apple or pear juice and add back a

R.W. Hartel and AK. Hartel, *Candy Bites*, DOI 10.1007/978-1-4614-9383-9_41, © Springer Science+Business Media New York 2014

raspberry flavor. It's more efficient and less expensive that way (apple and pear are the most abundant, and thus, cheapest, juices).

What's in juice concentrate anyway? A 1996 study by the National Food Processors Association (NFPA) analyzed 92 samples from all over the world over a period of three years to answer that question. Not surprisingly, the main components were sugars, with fructose contributing about 6 percent, glucose at about 2.5 percent and sucrose about 1.6 percent, when normalized to a standard single-strength juice concentrate. Interestingly, these apple juice concentrates also contained about 0.4 percent of sorbitol, a naturally produced sugar alcohol (often used in sugar-free confections).

Fructose (or high fructose corn syrup) is not usually used in confections because it's a powerful humectant; when added at too high a level, it causes candy to become sticky. But since it's naturally present in fruit juices, it's present in fruit snacks. So the distribution of sugars in fruit snacks is slightly different than what you would find in a bag of Swedish Fish or Dots or orange slices. Does fruit juice concentrate have any other ingredients that would make it better for use in fruit snacks?

Yes, it does. According to that NFPA study, the apple juice concentrates contained numerous minerals and polyphenols, a group of compounds thought to have substantial health benefits. Undoubtedly, the clarification and concentration processes reduced these levels substantially compared to the whole fruit (especially since these components are concentrated in the skin), but at least there's still some there in the concentrate. Is that enough to make a health claim for fruit snacks? You be the judge, but clearly corporate marketing people have no compunction about making that claim. And based on the number of fruit snacks sold, many consumers are OK with it as well.

Now let's compare the other ingredients in fruit snacks with those we find in jelly candies. If you recall (see Chap. 35), gummy and jelly candies contain a gelling agent, usually either starch, pectin or gelatin, although other gelling agents may be used to provide distinct textures. Both fruit snacks listed above contain modified starch as the primary thickener, exactly the same as

Dots and Swedish Fish. What's different is that these fruit snacks also contain other stabilizers to modify the texture of starch. The noncandy company added fruit pectin while the candy company used gelatin.

Mixing starch and either pectin or gelatin produces a texture somewhere between that of the two components. Pectin softens the texture of starch while addition of gelatin would increase the elasticity. There are also numerous confections that mix stabilizers to create unique textures. For example, Lifesaver Gummies have starch added to gelatin to make them less gummy (elastic) than a traditional gummy bear. So the mixture of stabilizers really doesn't distinguish fruit snacks from candy.

There are also numerous other candies that can, and sometimes do, make a health claim of sorts. Any candy made with peanuts or other nuts, like Snickers, PayDay or Peanut M&M's, can claim the health benefits of nuts (although it's interesting that nuts were on fire as being bad for us not much more than 10–15 years ago). How about candies with fruit bits, like chocolate covered raisins? Raisins are good for us, right, yet most still consider Raisinettes to be candy. Is it a healthier candy if it's acai fruit that's covered in chocolate?

And how about granola bars and other "sports or energy" bars? Although they typically contain fruits, nuts and grains, they still contain a fairly high sugar content. The binder that holds all the pieces together is made of sugar and corn syrup, the two main ingredients of confections. And they're often coated in chocolate.

The point to make here is that sometimes the line between what's a candy and what's a healthy snack has been blurred. As consumers, we make food choices on a daily basis. It's critical to your health and physiological well being that you choose fruits and vegetables on a regular basis, but it's also OK to enjoy candy, and fruit snacks, periodically and in moderation.

42

Sour Patch Candy

What's the sourest candy you've ever eaten? Debates abound on the internet; some favor Sour Patch candy while others say Warheads cause the most puckering. To my mind, Toxic Waste candy should be at the top of the list, if for nothing else than the name and the challenge. The sour challenge is written right on the container—"How long can you keep the candy in your mouth without spitting it out?" If you can keep it in your mouth for a whole minute, you're deemed a "Full Toxie Head!" Otherwise you're a "cry baby" or a "toxie wannabe." The sour gauntlet has been laid down.

What causes that sour sensation? The acid, of course. But what is an acid anyway? Although there are multiple definitions, for our purposes it's any molecule that can give up a proton to another molecule. When dissolved in water, an acid gives up a proton, in the form of a hydrogen ion (or more correctly, hydronium ion)—it's an ion because it has a charge, in this case a positive charge. An acid also turns blue litmus paper red and has a sour taste. Acidity is characterized by pH, essentially a measure of the number of hydrogen ions in water. Neutral pH, neither acid nor base, is 7.0. Things that are more acidic have lower pH values. Because pH is a logarithmic scale, a pH of 1 is incredibly acidic and, hence, can be incredibly sour.

Sour is one of the five known taste sensations, the others being sweet, salty, bitter, and umami (sometimes called savory). Although the details are still not completely understood, the sour sensation arises, at least in part, when hydrogen ions enter the taste buds on the tongue and neurotransmitters signal that characteristic sensation of sourness to our brain. The intensity of sourness is

R.W. Hartel and AK. Hartel, *Candy Bites*, DOI 10.1007/978-1-4614-9383-9_42,

proportional to the number of hydrogen ions, meaning that it correlates with pH. Candy makers play with the acid level to lower pH and dictate sourness of their products.

In the natural food world, fruits are the most common acidic products. For example, citric acid is found in citrus fruit, malic acid in apples, and tartaric acid in grapes, although most fruits have more than one form of acid present. For example, cranberries, widely recognized as being very tart, contain citric, malic, quinic, and L-ascorbic acids.

In candies, it's primarily citric, tartaric and malic acids that are used to provide sourness, although sometimes other acids (adipic, fumaric, etc.) can be found. A little bit of acid helps highlight fruit flavors, whereas a lot of acid generates a seriously sour sensation.

However, not all organic acids are created equal. Some are much more sour than others at equivalent use levels. In part, that depends on how many protons the molecule can give up; citric acid, in particular, is a triprotic acid, capable of giving up three protons, depending on pH. That makes citric acid one of sourest in the candy maker's arsenal. Also important is at what pH the acid gives up its proton(s). Lactic acid is only monoprotic (gives up one proton), but it does that at very low pH, 3.86, meaning it's also pretty sour.

And each acid has its own response on our senses. Citric and tartaric acids are very tart and provide an immediate burst of sourness, whereas malic acid has a smooth tart taste that builds more slowly, without a burst. Multiple acids are often used in concert to provide a sour crescendo, providing both the initial tartness intensity and the longer-term sour sensation.

Acid use in candies can be somewhat risky though. Besides the tart taste, acids can also wreak havoc with the sugars. In particular, sucrose is prone to break down in the presence of acid, particularly at elevated temperatures. The combination of sugar and acid in candies cooked to high temperatures induces inversion, or break-down of sucrose into glucose and fructose (see Chap. 12). The creation of glucose and fructose means that the candy will be very hygroscopic and sticky, with reduced shelf life. For this reason, acids are usually added to the sugar batch after it's been cooked, as it's cooling down prior to forming. This minimizes sucrose inversion and maintains desired quality characteristics. In fact, many candy makers use dextrose for sour candies because it's much more stable to acid than sucrose.

In case you were wondering, dentists typically don't have anything good to say about sour candy. The Minnesota Dental Association has put out a report on how bad sour candy is for your teeth. They compared the pH of sour candies against battery fluid, which has a pH of 1. Loss of tooth enamel starts at a pH of 4 and it proceeds faster at lower pH. The longer you keep a sour piece of candy in your mouth, the worse it is for your enamel and the greater likelihood of getting cavities.

In this document, they report the pH value of various candies, with lower pH being more acidic and, hence, more sour. The list starts with Spree with a pH of 3.0, which is also the value for SweeTarts and X-treme Airheads. At pH 2.5, the list includes Sour Punch Straws and Skittles. Starburst and Sweetarts Shock have a pH of 2.4, as do LemonHeads and Mentos. Two powder candies, Pixy Stix and Fun Dip have pH of 1.9 and 1.8, respectively. The candy on their list with the lowest pH? WarHeads Sour Spray, with a pH of 1.6, only a little less acidic than battery fluid. Unfortunately, they didn't have Toxic Waste candy on their list so we don't know for sure where it fits on the scale.

The litmus test for a sour candy is whether kids like it or not. Sour Patch Kids, with both tartaric and citric acids, enjoy huge popularity with kids of all ages. A starch-based jelly candy, Sour Patch Kids have acid within the candy but get the really sour punch from citric acid. That's not just sugar sanding on the exterior, that's a citric acid powder that gives the immediate sour kick. But it's not the sourest candy in the Sour Patch line. Sour Patch Kids Extreme takes it one step further, adding lactic acid to the tartaric and citric already present to really step up the sour punch. Based on the acid ingredients, Sour Patch Kids Extremes give the sourest punch for the buck.

43

Where Do the Jelly Beans in the Easter Basket Come from?

One of the most abundant items the Easter Bunny leaves in your basket each year is the jelly bean. According to the National Confectioners Association, 15 billion jelly beans are made each year in the United States, enough to fill a 9-story office building (really, who does all these calculations?).

Where do jelly beans come from? Some say that they're Easter Bunny droppings and I have a toy hen in my office that lays jelly beans when you push it to prove it. Actually, we know they're made in candy factories. The largest production facilities make thousands of pounds of jelly beans each day to fill our needs.

Jelly beans have been around a long time, with the first reference to them during the Civil War. Story has it that one manufacturer encouraged customers to send his jelly beans to the soldiers off at war. Their popularity grew rapidly in the early twentieth century and it's been increasing ever since.

Here's a fun experiment. Carefully dissect a jelly bean, any kind will do, and you'll see that a jelly bean has a jelly candy core and a sugar shell. There's actually another layer, but the polish on the outside is too thin to see.

The jelly bean center, a starch-based candy similar to a candy orange slice or gum drop (see Chap. 37), is made by cooking a slurry of corn starch granules, the dense energy source of the corn plant, with sugar and corn syrup. The heat and moisture cause the starch granule to swell and gelatinize into a thick syrup, similar to what happens when you make gravy, but much thicker. What makes this different is how much starch is added. A little starch gives a viscous liquid; more starch gives a gel structure.

R.W. Hartel and AK. Hartel, *Candy Bites*, DOI 10.1007/978-1-4614-9383-9_43,
© Springer Science+Business Media New York 2014

The hot candy syrup is poured into a bean-shaped mold impressed into dried powdered corn starch in a mogul tray (see Chap. 36). If you could look carefully at a jelly bean center, without the sugar coating (which hides that flat side), you'd see that one surface is flat—that was the top of the mold. While curing in the trays overnight, the dried starch removes water from the candy, allowing it to set into a firm jelly bean. The next day, the jelly bean centers are removed from the mold, sugar sanded to provide an adhesive surface and to prevent them from sticking together before being coated with sugar shells.

The second part of the jelly bean is the sugar shell. Jelly beans are called "panned" candies, because of the process by which the sugar shell is applied to the jelly center. Jelly beans are one of the most common "soft panned" candies, meaning that the sugar shell is easy to bite through, unlike the "hard panned" shell of an M&M or Jordan almond (see Chap. 45).

Large tulip-shaped pans, somewhat like cement mixers, rotate slowly with jelly bean centers tumbling inside. A sugar syrup, made of sucrose and corn syrup with color and flavor added, is poured into the pan to wet the surface of the tumbling beans. Once the syrup has uniformly spread on the surface of the jelly bean centers, a dose of confectioner's powdered sugar is applied. As the beans tumble in the pan, the forces applied as they land on each other cause the sugar crystals to pack tightly into the syrup to form the first layer of the shell. The process is repeated, up to five or even ten times, to get the desired shell thickness.

Getting the moisture content of each layer right is one of the key elements in sugar panning. One of the skills required to make good jelly beans is patience (see Chap. 45). Knowing when to just let the centers tumble in the pan to allow proper moisture migration is important for producing high-quality candies. Rushing the process often results in bags of clumped jelly beans as the water equilibrates later during distribution.

A soft-panned sugar shell has a structure somewhat analogous to cement: numerous small particles (but sugar crystals instead of gravel) held together in a matrix (the sugar syrup). The sugar syrup effectively acts like a glue to hold the crystals together. Unlike cement, however, the jelly bean sugar shell is soft and easy to bite through. The shell fractures at the point of the large crystals when your teeth bite through it.

The final layer on the shell is a polish, or a confectioner's glaze as it's often called in the ingredient list. The glaze gives the jelly bean its shiny appearance and protects it from damage in the package.

Confectioner's glaze is a euphemism of sorts though. It's actually an edible shellac, derived from a secretion of the lac bug. Shellac is a complex, and often quite variable, natural material. It's composed primarily of hydroxy fatty acid esters and sesquiterpene acid esters with a melting point somewhere above 70 °C (depending on source and purity). It's often found dissolved in alcohol although water-based versions are also available. The shellac used on jelly beans is food grade, generally recognized as safe (called GRAS by the FDA). It polishes your jelly beans just like it polishes your wood tables (but that's not food grade).

All in all, it takes anywhere from four to ten days to make a jelly bean, primarily to allow moisture equilibration between steps. After the centers are deposited into the dried starch, they can take one to two days to completely set. After being removed from the starch, they're steamed and sanded with a granular sugar coating. Once that coating dries, they go into the pans to apply the sugar shell. That shell has to set at least overnight prior to polishing. After the polish has set, they can finally be packaged.

Besides being delicious to eat, jelly beans find numerous ways to sweeten our lives. They make good replacement chips for penny ante poker. They can be used to make bracelets for five-year old girls. They can be filled into a jar for a "guess the number" raffle. And they've been used to create jelly bean art.

Probably one of the more creative uses for jelly beans has to be the performance art piece using jelly beans to mark the typical life. Artist Ze Frank compiled 28,835 jelly beans to represent the number of days in an average life span. He then separated them out according to days eating, sleeping, working, watching TV, and so on. The final conclusion is that there are less than 3,000 jelly beans left in your entire life for doing whatever you want—your hobbies or whatever. Seems like a meager number of jelly beans for free time, but that's actually over eight years of your life. We thank you for spending a few jelly beans of your life with our book.

44

Jelly Bean Flavor Development

Wouldn't it be cool to work as a product developer for Jelly Belly? Coming up with the latest fun flavor sounds like an ideal job. Well, maybe, maybe not.

Dog Food. Centipede. Moldy Cheese. Baby Wipes. Imagine working as the jelly bean developer for these flavors. Yuck.

We asked a confectionery flavorist how you go about developing a flavor like Vomit. How do you match the flavor of barf? Do you have to collect samples and run analytical tests before putting together the specific compounds that give the flavor? She admitted that she was the person who developed that particular flavor, but unfortunately couldn't share the specific steps she went through to reach the final commercial candy.

Typically, the process goes like this. First, someone at Jelly Belly comes up with an idea for a new product flavor that the bosses think has potential to be a big seller. Often that's related to some pop culture fad.

For example, back when Harry Potter was the rage, the idea of marketing weird candy flavors was a no-brainer. The popularity of the Harry Potter movies, including the odd flavors of Bertie Botts, would carry over into Jelly Belly beans with those weird flavors. Co-branding Harry Potter and Jelly Belly was a smart business move for both sides.

It's common for a company like Jelly Belly to approach a flavor company, specialists in putting together the specific chemical compounds that invoke a specific nuance, to help develop complex flavors. There are numerous such companies willing to help develop new and novel flavors, each claiming a special expertise. Being the

R.W. Hartel and AK. Hartel, *Candy Bites*, DOI 10.1007/978-1-4614-9383-9_44, © Springer Science+Business Media New York 2014

flavor developer for a really big commercial hit means good profits and job security for the particular developer. That means there's a lot of competition.

Flavors are either natural or synthetic. As you might expect, natural flavors need to come directly from some natural source, with a limited amount of processing to turn it into a form that can be added to a jelly bean. The most common natural flavors come from fruits, usually in the form of concentrates. Another natural flavor is vanilla since it's extracted directly from the vanilla bean.

Can you imagine a natural source for vomit flavor? No! For sure, this one's going to be synthetic.

And in fact, most flavors are synthetic, a combination of specific flavoring compounds contained in some sort of carrier. Commercial flavors contain a small percentage of the actual flavor molecules in a liquid carrier—water, alcohol, oil or some other organic solvent, depending on the nature of the flavor molecules. Some flavor compounds prefer water (hydrophilic) while others prefer organic solvents (hydrophobic). It's those flavor components that are most important, although sometimes the carrier can influence the candy product in other ways (see Chap. 57).

Flavor molecules are highly volatile by definition; they prefer to be in the vapor (gas-like) state at room temperature. Controlling flavor release from the candy is a key element of any candy product design. We want to some of the flavor to release when we smell the candy but we want most to be released when we eat it.

During candy manufacture, the flavors are mixed into and embedded within the candy matrix and hopefully are only released when we bite or suck on the candy. In gourmet jelly beans, the flavors are contained in both the sugar shell and the jelly candy center. That's been Jelly Belly's main claim to fame, although there have been numerous copycats. Prior to that, jelly beans only contained flavor in the sugar shell—the center was unflavored.

When you bite into a Jelly Belly and crack open the shell, those imbedded volatile flavors are released into your mouth and on up into your nasal cavity. The "flavor" sensation is primarily an aroma, where the flavor molecules interact with your smell sensors in the nose. When you chew a Vomit Jelly Belly, the flavor of upchuck has to release into your nose to give that whiff of fresh spew.

For most flavors, there are a few chemical compounds that constitute the majority of the flavor. Isoamyl acetate is the primary compound in banana flavor, benzyl acetate is strawberry, and limonene is the primary flavor constituent of lemon. Other flavors are much more complex. Chocolate, for example, contains hundreds of different flavor compounds, each of which adds subtlety and nuances. This is why there are no good synthetic chocolate flavors.

To develop the Vomit flavor, a flavorist needs to understand which chemical compounds they need to blend together to invoke that special aroma that comes from losing your lunch. How do you do that? First you have to analyze a sample. Collect a sample of hurl and send it through the gas chromatograph to see what flavor compounds predominate. However, from my experience, the "flavor" depends on what I ate for lunch before it came back up again. I guess the flavorist's job is to find the most average hurl flavor possible.

As with development of most flavors, the flavorist would identify those primary components, put them together in a base solvent, and test it out on someone. That means smelling the flavor and evaluating it in some vehicle (like a candy). If the first attempt doesn't quite match the target puke aroma, you go back and adjust the chemical profile again. For each iteration, someone has to smell the upchuck and maybe even taste test it in a model system to decide if it meets the customer's needs. Once the flavorist is satisfied with a prototype, the flavor is then incorporated into a jelly bean and tested with a sensory panel.

In food companies, sensory panels are often made up of employees. They're used for a variety of tests, but often would be used to evaluate things like new flavors. When the prototype flavor is ready, some test jelly beans with slightly different versions of vomit flavor are made for the sensory panel to evaluate. The beans are put in front of the panelists, who then taste and rate each jelly bean variation. Sometimes numerous iterations are required, especially with complex flavors like gut soup.

The job of the sensory panelist is a mixed bag. With fun flavors like Root Beer and Pina Colada, the job is enjoyable. But with flavors like Vomit and Dirt, it's a not a job for the faint-hearted.

Does designing a flavor for Jelly Belly still sound like a fun job? Although developing a nasty Vomit flavor may not be so pleasant, in general, new candy flavor development is a fun challenge.

45

Panning Patience

What do teaching and panning (putting a sugar shell on a nut or chocolate lentil) have in common? Patience. They both require lots of it.

As a graduate teaching assistant (TA) many years ago, I found out the hard way how much patience is required for teaching. I was in charge of a self-paced modular Physics lab, trying to help people understand the nature of the world around them through physical concepts. To me, it was obvious, but to some of the students, it was worse than learning Latin would be for me. I found out the hard way that I didn't have the patience needed and swore I'd never be a teacher.

Now that I'm a candy scientist, I've also come to know how difficult panning is. Panning is the process of sequentially building up layers of sugar or chocolate coating on to a candy center. Examples of sugar panned products with a soft shell include jelly beans (see Chap. 43) and Lemon Heads. Examples of sugar-panned candies with a hard shell include M&Ms, gumballs, and Jordan almonds, while chocolate covered raisins, peanuts and almonds as well as malted milk balls are chocolate panned products. While all panning requires patience, hard panning arguably requires the most.

Panning got it's start hundreds of years ago as candy makers learned to control sugar and it's different states. By tossing nut centers into melting sugar in a pan over a fire, they were able to make a sugar shell similar to that on a Jordan almond.

Traditional hard panning nowadays means standing in front of a pan in which a sugar shell is being applied to tumbling centers. A

R.W. Hartel and AK. Hartel, *Candy Bites*, DOI 10.1007/978-1-4614-9383-9_45,   179 pan is a rotating drum that's often tulip-shaped. The tulip shape design controls the tumbling movement from front to back as well as up and down. Alternatively, pans may be spherical in shape, or more accurately, an oblate spheroid, like the earth. Imagine a hollowed-out earth rotating on its axis with the North Pole cut off (sorry Santa) filled with candy centers awaiting their sugary coating.

Doses of sugar syrup are sequentially sprayed or ladled onto the candy pieces as they roll in the pan to build a layer of appropriate thickness and consistency. It's critical to make sure each dose of sugar syrup crystallizes and dries completely before adding the next dose. Panning involves knowing exactly when and how much of the next layer to add. It often means twiddling your thumbs while waiting rather than forging ahead with the next layer. Patience is a virtue in panning.

The same goes for teaching. A good teacher patiently waits for a student to grasp a concept before moving on to the next teaching point. In doing so, student learning develops from a solid foundation.

What happens if either the teacher or the panner loses patience? In teaching, the student gets pushed beyond the capability to understand a concept and incorporate new material into what's already known. Or worse, sees the frustration in the instructor and either loses interest or feels belittled. Although I was completely frustrated when my Physics students couldn't grasp the topic, I'm sure they were equally, if not more, frustrated by my inability to put things into a form they could grasp.

In panning, particularly panning a hard sugar shell, losing patience can be a disaster of major proportions, as we'll see.

M&Ms are one of the most notable of hard panned candies. They first came out in 1941 and were named after the two people who developed them, Forrest Mars Sr. and Bruce Murrie (who had ties to the Hershey Company). Supposedly Mars had seen a chocolate-coated type product during the Spanish Civil War and came up with the idea of sugar coating a chocolate center to help protect the chocolate from the heat. Hence, the slogan "Melts in your mouth, not in your hands."

Arguably one of the first hard panned confections, Jordan almonds, have been known for centuries. They're sometimes called sugar almonds, or the generic term dragée (a bite-sized candy with a hard sugar shell). Jordan almonds have become synonymous with weddings, and other celebrations. A wedding tradition is to give five Jordan almonds as favors, each representing a different wish—health, wealth, longevity, fertility and happiness.

Both Jordan almonds and M&Ms, as well as many other hard panned products, are made in large rotating pans. The centers tumble in the pan as it turns, with multiple doses of sugar syrup sprayed on to build up sequential layers. Each layer is only about a tenth as thick as a human hair, so even to build a shell a couple millimeters thick requires multiple coats. Some hard-panned candies have well over 30 separate sugar applications to build up the shell.

After each dose of sugar is applied to create a layer, a period of time is required before the next dose can be applied. During this quiet time (if any panning operation could be called quiet; imagine hundreds of nut centers tumbling in a metal pan—the noise is deafening), several things are happening. For one, the water in the syrup that's been applied starts to evaporate, particularly since dry air is applied to the pan after each dose. At the same time as the water evaporates, the sugar in the syrup begins to crystallize. The brittleness of the hard sugar shell arises from numerous small sugar crystals that are partially fused together to form a network so it's important to make sure the sugar completely crystallizes.

But crystallization is much slower than drying. To dry the thin layer of sugar syrup in a pan takes much less than a minute, whereas crystallization takes place much slower, on the order of five to ten minutes under these conditions (and probably hours to completely equilibrate). Those sugar molecules take time to coordinate and organize into a crystal structure. And drying too fast actually slows down crystallization even more. It's critical to allow sufficient time for crystallization prior to addition of the next syrup dose. The panner essentially must twiddle his thumbs while waiting for crystallization to be complete.

What happens if the next layer is applied too soon? Crystallization continues but more slowly, and any water trapped below subsequent layers eventually will work it's way out, carrying soluble color molecules along with it. Over the next few days after manufacture, the color of a hard panned candy shell that was processed too quickly becomes mottled, with white spots where color has left with the water. Instead of a shiny vibrant color, the candy appears dull and splotchy.

Candy makers try to minimize these changes and speed the process by adding colors that are "tied" to a particle. We call these "lakes," where a dye molecule is tightly attached to a microscopic aluminate particle. Dye molecules on a lake don't move, even when water molecules are moving around, so the color stays uniform. Unfortunately, lakes don't give as good a color as soluble dyes, so a mixture of the two is generally used.

The other solution to the mottling problem is to slow down, practice some patience. Although slowing down production flies in the face of modern manufacturing practices, practicing a little patience now pays dividends for the future in product quality.

The same is true in teaching. The best learning comes when the teacher slows down to make sure the student has a good grasp of the concepts before building to the next level. But teachers often face the same type of pressure as the manufacturing plant; we have to cover a certain amount of material in the syllabus to prepare students for the next class.

46

Everlasting Gobstoppers
and Atomic Fireballs

Gobstoppers have been known in England for nearly a century, with the name originating from the word gob, slang for the mouth. Hence, gobstopper—something that stops the mouth. If you want to stop someone from prattling on, simply stuff a gobstopper in his or her mouth.

The Everlasting Gobstopper, a jawbreaker that changes colors and flavors, was the brainchild of Roald Dahl in *Charlie and the Chocolate Factory* in 1964. They were intended as poor kids candy since they magically kept regenerating, no matter how long you sucked on them. A subsidiary of Quaker Oats, Breaker Confections, began making them to cash in on the success of the movie, Willie Wonka and the Chocolate Factory, the version with Gene Wilder dancing and tumbling in the title role. The Everlasting Gobstopper is now produced by Nestle through the Wonka brand, along with numerous other fun candies like Nerds and Runts (see Chap. 47).

Unfortunately, the everlasting part is just fiction. A real gobstopper only lasts for so long. Still, the gobstopper lasts a good long time, longer than most other candies, with the possible exception of the all-day sucker. It also provides extra entertainment value as the colors and flavors change.

In fact, a gobstopper is built by adding sequential layers of candy onto previous layers, in much the same way that a tree grows by adding rings of woody material to previous year's growth. Look carefully at the cross-section of a gobstopper—you can read the multiple rings of different color and flavor like the rings of a tree

R.W. Hartel and AK. Hartel, *Candy Bites*, DOI 10.1007/978-1-4614-9383-9_46,
© Springer Science+Business Media New York 2014

trunk. As you suck through one layer, the color and flavor change once you start dissolving the next interior layer.

Along the same lines, it's actually quite informative to watch a gobstopper dissolve in water, and a fun experiment to do at home on a rainy day. Space different colored gobstoppers around the edges of a shallow pan of water and watch what unfolds. As the sugar in the gobstopper dissolves, the water takes on the color of that shell, with a distinct "front" of color spreading out in rings from the candy at the center. When two color fronts collide, there is a time when there is a sharp line of delineation between the two colors caused by slight differences in density. Over time, though, the color molecules of one line diffuse into the other to smear the color into some indistinct shade of brown.

What's happening is that the color, embedded within each crystalline sugar layer, diffuses out into the water as the sugar dissolves and spreads. However, there are two types of colors used in Everlasting Gobstoppers—a soluble dye and a lake. As the name implies, a soluble dye consists of color molecules dissolved in the water. As water molecules move, so do the color molecules. A color lake, however, is different. A lake is the colorist's term for a small particle that carries color. In a lake, dye molecules are adsorbed to a microscopic particle of alumina. They provide color through dispersion. Because these are small particles, they follow their own rules of migration, and it's these lakes that cause the temporary delineation between colors in the gobstopper dissolving experiment above.

Gobstoppers and jawbreakers are built layer after sugar layer in the hard panning process. Starting with a small bit of candy, sometimes even just a single sugar crystal, the layers are added step by step. First, a dose of sugar syrup is applied as the centers tumble in the rotating pan. Each sugar syrup application is followed by a waiting time when drying and crystallization occur. Each application of syrup takes several minutes to apply, crystallize, and dry. And it requires hundreds of layers to build up to gobstopper size. In fact, each ring of color seen in the cross-section is made of numerous individual sugar applications. Hence, it can take days build up a jawbreaker from scratch. The softball sized jawbreaker takes weeks to build up.

A hot cousin of the gobstopper is the Atomic Fireball, produced by the Ferrara Candy Company. They're made in the same sequential steps as any gobstopper—a dose of sugar syrup is applied to the pieces tumbling in a pan followed by a brief period of drying and crystallization. In essence, the Atomic Fireball is a hopped-up, cinnamon-flavored jawbreaker.

What causes Fireballs, and Cinnamon Red Hots, to be so hot? Cinnamon itself, used as a common spice, isn't hot like a chili. In French toast or cinnamon rolls, cinnamon provides an interesting flavor, but definitely not spicy hot. The bridge between cinnamon (cassia) flavor and spicy heat was part of the genius behind the original Atomic Fireballs developed by Nello Ferrara in 1954. It quickly resulted in huge success for the company. To get the spicy heat, the cassia flavor is spiked with capsaicin, the spicy compound in hot peppers. It's so hot that they keep the bottles of flavor double-wrapped with warning labels.

Some people find Fireballs a bit too spicy, while others relish the challenge of keeping six of them dissolving in their mouth without spitting them out or choking up. But there are stories of other, hotter candies. One candy, appropriately called Ghost Pepper Dragon Boogers, a gummy candy in the shape of a chili, contains the essence of the Bhut Jolokia chili pepper, making it about 200 times more potent than the Atomic Fireball. Would eating one Dragon Booger be like having 200 Atomic Fireballs going off in your mouth?

While Fireballs and Dragon Boogers provide some hot spice to life, Everlasting Gobstoppers have some heat sensitivity too. They respond in their own way to heat—by exploding. Apparently, when left in the hot sun for too long, or heated in the microwave by the Myth Buster crew, internal sugar layers heat faster than outside layers. When subsequently bit into, or crunched in a vise by the Myth Buster crew, hot molten sugar can spew out, burning unsuspecting candy eaters.

The Gobstopper would then be sort of an "atomic fireball," with serious health consequences. Please be careful with these "hot" candies.

47

Runts and Nerds

Runts and nerds, names a small, shy kid might be called by the bullies in grade school. How did they get to be names of famous candy brands?

Both candies were developed in the early 1980s by the Willie Wonka Candy Company, with Runts coming out the year before Nerds. Both are now part of the Wonka division of Nestle. Runts are small, fruit-shaped candies with a hard shell and powdery center, while Nerds are even smaller, irregular-shaped hard candies (although not in the category of boiled sweets—see Chap. 13).

Perhaps Runts were named as such for their diminutive size. Then when they came out with Nerds the next year, even though they're significantly smaller, the name was already taken. They must have decided that Nerds were cousins of Runts and went with that.

Both Runts and Nerds are hard panned candies (see Chap. 45), but with several differences. First, the centers that start the panning process are different. Runts use a compressed tablet as the starting point, similar to SweeTarts, another Wonka candy. The give-away in the ingredient list is calcium stearate, the compound used as a lubricant in the tablet press (see Chap. 20). The tablets are pressed in the specified shape; for example, the banana-flavored Runt is pressed into the long arc representative of a banana. The other shapes are less specific to the flavor, with the red cherry Runt in the shape of a heart. These pressed tablets are panned to apply the hard sugar shell that provides the crunch characteristic of Runts. Once you break the hard sugar shell, the softer powdered tablet in the center is released.

R.W. Hartel and AK. Hartel, *Candy Bites*, DOI 10.1007/978-1-4614-9383-9_47,
© Springer Science+Business Media New York 2014

Nerds are even simpler. A single sugar crystal serves as the center for creating a hard shell. As one candy-maker said, a 50-pound bag of table sugar provides the starting point for an awful lot of Nerds. As hard-panned sugar crystals, Nerds are irregular-shaped, seriously crunchy candies.

Runts and Nerds are different from most other hard-panned candies. In both candies, the main ingredient isn't "sugar," a generic term that really means sucrose. Instead, the main ingredient is dextrose (also called glucose). In a Runt, the entire candy is made from dextrose, from the pressed tablet center to the hard sugar shell. In contrast, a Nerd is a dextrose-panned sucrose crystal.

Dextrose panning provides several unique differences from the more traditional sucrose panning (see Chap. 45). From a sensory standpoint, dextrose provides a significant cooling effect compared to sucrose. When you eat a Nerd or Runt, the dextrose dissolves and dilutes in your saliva and that change requires energy, called the heat of solution. That energy comes out of your mouth, giving a distinct cooling effect. What makes dextrose different is that it has a heat of solution that's six to seven times greater than sucrose. A sucrose crystal barely gives a noticeable cooling effect while dextrose causes a chill as it's released into saliva.

Another significant difference between sucrose and dextrose panning comes from the nature of each crystal. Sucrose crystallizes in an anhydrous form, meaning all water molecules are excluded from the organization of sucrose molecules into a crystal, whereas dextrose crystallizes as a monohydrate. Each molecule of glucose in the crystal is accompanied by one water molecule. Since hard panning involves syrup application followed by a period of drying and crystallization, drying in dextrose panning is significantly easier since some of the water actually remains in the shell, tied up with the crystals as they form. Dextrose molecules also crystallize rapidly in the panning process into many small crystals that help form a crunchy shell.

One additional difference between sucrose and dextrose panning relates to the flow properties of the syrup that's applied to the tumbling centers. The viscosity of this so-called engrossing syrup is important for successful panning. Too thin and it splatters all over the pan, but too thick and it doesn't coat the pieces tumbling in the pan. Like porridge and serving temperature, the viscosity of engrossing syrup needs to be just right. Dextrose, because it's a monosaccharide, is much thinner (less viscous) than sucrose, a disaccharide (meaning two monosaccharides stuck together) at equivalent concentration. To candy makers, this means they can use a higher concentration engrossing syrup, so again there is less water to evaporate off in each syrup dose. Dextrose panning goes significantly faster than sucrose panning, with equivalent shell thickness built in much less time. To panners, time is money and dextrose panning allows more product out the door every hour. As one Wonka candy maker said, why would you pan with anything but dextrose?

Spree, another Wonka candy, is also a compressed tablet made of dextrose and coated with a dextrose shell, same as Runts, just with a plainer, disk shape. Spree was developed in the early 1970s by Sunline Confections, which was bought by Nestle in 1989. In that same acquisition, Wonka got several other well-known dextrose-based confections, including Pixy Stix, SweeTarts, and Fun Dip.

A recent addition to the Runts and Nerds candy collection is the Nerds Rope, a bunch of Nerds glued onto a gummy candy roll. Instead of being packaged freely, like normal Nerds, these guys are stuck in place and not going anywhere. They also provide a unique texture contrast—the crunchiness of Nerds with the elasticity of a gelatin gummy candy rope. Another interesting line extension is the Nerds gumball, a hollow sphere of bubble gum filled with crunchy Nerds. There was even a Nerds cereal at one point, but it's now in the dead cereal graveyard. I guess even the loosest parents couldn't justify serving their kids a cereal based on a popular candy.

From the wide array of spin-off products from Nerds and Runts, it's clear that Willy Wonka has been busy tinkering away in his candy lab. Maybe the next candy in the lineup will be Geeks. If you were Willy Wonka, what candy would you make?

48

Is Licorice Good for You?

What would King Tut do if he'd awoken one morning with a sore throat and a cough? Maybe he'd reach for a menthol cough drop? More likely he'd reach for a natural cure, like licorice.

One of mother nature's most prolific natural curative materials, licorice is not only a cough suppressant and throat soother, it's also been associated with easing the pain of ulcers and enhancing the body's resistance to stress. Its anti-inflammatory properties help soothe arthritis and some people claim it's effective against chronic fatigue symptom. And its laxative effect may help keep you regular. One recent study even claims that consumption of licorice reduces risk factors of cardiovascular disease.

Perhaps that's why licorice was discovered in King Tut's tomb and used by such historical luminaries as Alexander the Great, Caesar and the Indian prophet, Brahma. It's good for almost whatever ails you. And it's natural.

But before you run out for a big bag of licorice to cure all your ills, you should understand what it is (and what it's not), how it's made (see Chap. 49), and some caveats about eating too much of it.

Licorice is a plant (*Glycyrrhiza glabra*) indigenous to parts of Asia and the Mediterranean region, but is cultivated around the globe. In England, the town of Pontefract is the unofficial licorice headquarters, the center of licorice cultivation in Britain. Pontefract hosts a liquorice (the British spelling) festival each year, sporting licorice everything, from candy to cakes. Supposedly, they even make licorice cheese. Check it out to see who's elected the Liquorice Queen this year.

R.W. Hartel and AK. Hartel, *Candy Bites*, DOI 10.1007/978-1-4614-9383-9_48,   191
© Springer Science+Business Media New York 2014

The licorice plant is a tall shrub with blue/violet flowers, but it's the root that contains the ingredient for use in licorice candy.

The root contains numerous chemical compounds, many of which may infer the health effects noted above. But the compound most noted in the licorice plant, and the one responsible for its name, is glycyrhhyzin or glycyrhhizic acid. Since glycyrhhyzin is 50 times sweeter than sucrose, the licorice root is sometimes called the sweet root. In fact, the Greek word glykyrrhiza means exactly that, sweet root.

Our ancestors, supposedly dating back to 2737 BC China, probably started chewing this root and realized that not only was it sweet, but they could attribute to it all those wonderful health benefits. As the population grew and became more developed, chewing licorice roots wasn't real convenient (or tasty) and the roots probably couldn't be shipped very far without spoiling, so people looked for ways to preserve it. They probably dried the roots and shipped them without trouble, but extraction of the essences became the main way to preserve and transport licorice. After extraction, the juice was concentrated for efficient shipping and use.

This concentrated licorice root extract is a paste of sorts, which is then formed into a block. Block licorice, which is what licorice extract is sometimes called, contains glycyrhhizin of course, but it also contains concentrated bitter compounds that impart the characteristic licorice flavor. To turn this into something palatable, our ancestors figured out ways to use licorice extract to make all manner of interesting foods, with licorice candy being the number one choice.

It may seem logical that all licorice candy contains licorice extract, but think again. What we call licorice in the United States most often doesn't contain any licorice whatsoever. Since licorice extract is a dark brownish/black color, any candy made with licorice extract will take on that color and has a strong characteristic licorice flavor. That's black licorice.

The sweet red twist product characteristically called licorice by American consumers is not really licorice. I repeat, red "licorice" is NOT licorice. It doesn't even say licorice on the label. Twizzlers, the number one "licorice" product in the US, is simply called fruit-flavored twists on the label. Nowhere, especially not in the ingredient list, does it say licorice or licorice extract. True licorice comes in only one color—black. And it has that characteristic black licorice flavor, which by the way is often enhanced in commercial products by addition of anise, a licorice-like flavoring.

At what point does common usage of a word trump the original meaning? Ask any American school kid if red Twizzlers are licorice and absolutely every one of them will say yes. So how can we argue with all those kids being trained to call red fruit twists licorice?

As an aside, another example of a word that has been changed by common usage is toilet. For most all of us, the word connotes the porcelain unit in which we do our business or, by extension, the room in which that bowl resides. The word originates, however, from the French word for the cloth that was draped over the shoulders of a lady while her hair was being dressed.

Regardless of the name, if you want to enjoy the health benefits of licorice, you'll have to eat the black version. Since those red licorice fruit twists don't contain licorice extract, you don't get the benefits. Call it whatever you want, you still need to eat black licorice to get the health kicks.

But wait, before you run out for that bag of black licorice to help ward off life's woes, you need to listen to the potential side effects. Reminiscent of any drug commercial, this chapter concludes with the mandatory *ad nauseum* list of potential side effects of licorice. Read the next few sentences as fast as you can to emulate one of those annoying drug commercials where it's now mandatory to conclude by giving all the potential side effects. *Do not eat licorice if you have high blood pressure or hypokalemia edema. Licorice may cause posterior reversible encephalopathy syndrome. Do not eat licorice if you have certain liver disorders or diabetes. Do not eat licorice if you're pregnant. Some people may experience water retention, upper abdominal pain, headache, shortness of breath, and stiffness after eating licorice.* And the list goes on, but you get the idea.

49

Licorice Variations

Although we know that licorice is really a flavor (see Chap. 48), it's come to be known as a separate candy category of its own. The variety of licorice-type products available in the candy store is truly huge. It comes in different colors and flavors and a wide array of different shapes. Licorice candy can be found in sandwiched layers or filled tubes, it can be in pellet form or it can be covered in a sugar shell. You can buy licorice in long skinny ropes, as rolled up bands, in shoe-strings, as nubby nubs, and as wound round strands to pull and peel. Licorice manufacturers have been very good at inventing new shapes and styles of licorice to appeal to our senses of sight, taste and touch.

Licorice is one of those products that begs to be played with. In fact, licorice manufacturers seem to go out of their way to develop products that appeal to the kid in us. The pull and peel product is a good example. It's nothing more than a bunch of single strands of licorice that are bundled around together with a slight twist. To eat, you can either bite off huge chunks of the bundle or take your time and peel off individual strands from the bundle to eat more slowly.

By the way, did you know that Twizzlers Pull n' Peel provides the perfect way to learn how to braid hair. In three bundles of three, it's the perfect combination for practicing braiding—and then you can eat it when you're done.

Licorice ropes are fun to play with too. A nice childhood memory is buying the shoestring licorice at the park and slowly eating my way from one end to the other while playing game after game of Nok Hockey (a stick and puck game common to the New

York area). A three-foot long strand of red licorice could last all morning.

You can also have a knot-tying session using licorice ropes. It's a good way for kids to learn how to tie knots and again, they can eat their work. Parents, you might consider this a rainy day activity for young kids—it combines an educational experience, assuming someone knows how to tie knots, with an afternoon treat. Bring grandma along to help with the granny knot.

How do licorice makers produce this wide array of products? In an extruder—the primary step in the licorice manufacturing process. After the candy paste has been cooked, the mass enters an extruder to be formed into whatever shape is being created that day on the licorice line.

Technically, an extruder is a machine that forces ductile or semi-soft materials through a hole (or die) under pressure to create a specific form or shape. Extruders themselves come in a wide variety of types. One of the simplest extruders is the cookie press, a cylinder filled with semi-soft cookie dough with a piston to press the dough through some sort of shape. Can you visualize the shape of the holes needed to form a Christmas tree cookie shape? The die hole doesn't always look exactly like the piece after the dough passes through (the reason is complex, related to the flow properties of the dough).

A pasta press is another simple example of an extruder. Wet pasta dough is forced through a small die hole. The pieces of pasta can be cut short as soon as the dough passes through the die or long ropes of pasta can be collected on a conveyor to pass through a drying oven. There are almost as many different shapes of pasta as there are for licorice, thanks to the extruder.

A licorice extruder is similar to a pasta extruder in many respects although it's a bit more complicated. It's got two screws that rotate together to drag the dough from the inlet hopper to the outlet at the die hole. The two screws form a channel through which the licorice paste flows. The pressure builds up inside the extruder as the dough is forced against the die plate at the end of the extruder.

That pressure squeezes the licorice paste out of the die holes, to be collected by take-away conveyors.

The shape of the die determines the shape of the licorice. A simple cylindrical die hole makes a continuous rope of licorice, like those strings I used to eat at the park. A cylindrical die hole with a second cylinder in the middle makes the hollow ropes of licorice, and a helical groove on the outside of the hole creates the popular twist shape (according to Guinness, the longest twist licorice extended for 1,200 feet and weighed 100 pounds!). To keep the hole in the center of the licorice rope, air is blown through the inner die hole at the same time the licorice exits the die. To make the pull and peel twist candy, numerous ropes of licorice are extruded continuously through a multi-holed die plate, with the strands bundled together and the entire bundle twisted as it exits the die plate.

Licorice allsorts, licorice tubes and licorice sandwiches filled with sugar paste candy, are another unique licorice candy creation. Licorice allsorts supposedly originated in 1899 at the Bassett's company in England when a sales person dropped sample boxes of the individual different candies (chips, rocks, buttons, nuggets, plugs and twists) onto a counter and the client became intrigued enough with the mixture to place an order.

Sandwich-shaped allsorts containing alternate layers of flavored sugar paste and black licorice used to be made by hand, with each layer applied separately prior to cutting the pieces. Now, they're made by co-extruding the multiple layers into a slab on a conveyor and then cutting the slab into the appropriate sizes. Cylindrical allsorts, also co-extruded, are either a tubular center of black licorice surrounded by a layer of flavored candy paste or the paste is inside with the licorice layer outside.

To enhance the variety of allsorts, both licorice and sugar paste flavors can be interchanged. Coconut is a popular sugar paste flavor, but chocolate and vanilla find their way into the bag as well.

By the way, Twizzlers licorice, the number one brand in the United States, has been around for a very long time. It's one of the oldest brands of candy in America. The Y&S (Young and Smylie)

Candy Company, founded in 1845 in Brooklyn, NY, claims to have started the Twizzlers brand. Twizzlers licorice is now produced by Hershey.

The number two licorice producer in the United States is Red Vines, a product of the American Licorice Company. They claim to produce over 115 million pounds of product each year, most of it the red variety.

Numerous other varieties are available as well, including products from as far away as Finland and Australia. In fact, licorice has an international appeal, with some countries eating way more licorice than us Americans. In countries like Holland, Denmark, Iceland, and other Scandinavian/northern European countries, the average person eats up to 4.5 pounds of licorice per year, a lot more than Americans eat. And it's not the same licorice. It's black, it's hard and it's salty, and it's definitely an acquired taste.

50

The Marsh Mallow

Where do marshmallows come from? There is actually a plant called the marsh mallow, *Althaea officinalis*, from which the original marshmallow was derived and to which its name is attached. However, the modern confection is quite different and no longer relies on the marsh mallow plant.

The marsh mallow plant has long been known for its medicinal properties and, in fact, that was the purpose of the first marshmallow-type confection. Different parts of the plant provide different health or medical benefits, with the root having been used since medieval times for soothing a sore throat. Some French pharmacists extracted the juice from the marsh mallow plant roots, cooked it with added sugar and egg whites, and then whipped the concoction into a meringue-like confection. When dried, it helped ease the sore throats of children, although adults most likely ate it too. At some later date, confectioners changed from the marsh mallow extract to gelatin as the stabilizer to hold the air bubbles generated by whipping. Perhaps the switch was made because gelatin gave a better, more consistent product or perhaps because it made marshmallow manufacturing easier.

Modern marshmallow manufacture is now highly automated and has been since the early 1950s when the current process was first developed. Numerous improvements and advancements now allow production of thousands of pounds of marshmallow a day. That's saying something because most marshmallows contain more air than anything else.

With a specific gravity (ratio of density of marshmallow to density of water) of 0.3–0.4, or even lower for some highly aerated products, marshmallows are more than half air. It's a confectioner's dream, to sell more air than candy—air is easily the cheapest ingredient in a marshmallow, it's free.

What holds all that air in a marshmallow? Gelatin. Although marshmallow-like products can be made with other proteins, gelatin is what gives the elastic bounce of the classic marshmallow. Gelatin, a breakdown product of collagen, stabilizes the air bubbles in marshmallow. It's also the primary ingredient responsible for gummy bear texture (see Chap. 39), although it's used in much smaller amounts in marshmallow. That's why marshmallows are easier to eat than gummy bears.

The process for making marshmallow is slightly different if you're doing it at home than in the manufacturing plant. At home, a mixture of corn syrup and sugar is boiled to about 227 °F to give a moisture content of 20 percent or so. In a separate step, gelatin is hydrated with enough warm water to make a thick solution. Once the sugar syrup has cooled to about 100 °F, the gelatin solution is blended in along with any desired flavoring and whipped in a Kitchen Aid or Hobart-type mixer to reach the final density. The marshmallow is then scooped out of the bowl, slabbed on a table, and cut into pieces for serving. Because of the high water content, marshmallow is incredibly sticky. That's why it's often coated with either sugar or starch.

In the home process, the gelatin is added after the syrup has cooled down because it's sensitive to heat degradation. In the same way that gristle in meat cooked in a hot crock becomes tender over time, gelatin breaks down at elevated temperatures, and at a faster rate at higher temperatures. To ensure no degradation of the gelatin, it's not cooked with the sugar syrup. It has to be added later.

In commercial marshmallow manufacture, the entire process is streamlined and fully automated. From mixing the syrup to packaging the finished product, the entire process is overseen by one operator, a technician familiar with machine operation. As long as the machine runs correctly, no knowledge of marshmallow science is needed.

In commercial operations, the gelatin is simply cooked with the sugar syrup, rather than being added later after the syrup had cooled. In this case, kinetics play an important role, with both time and temperature factoring in. If the gelatin was added at the beginning of a batch that was then cooked to 235–240 °F in 20–30 minutes, a significant amount of gelatin would break down. The marshmallow would have reduced springiness from that loss of gelatin. But since the time the syrup spends at elevated temperature in modern cookers is so short, there is little to no degradation of the gelatin. It's simply easier to dump it in at the start.

After the gelatin-containing syrup is cooked, it's allowed to cool a bit before being aerated. Whipping is generally accomplished in a rotor-stator type device. Compressed air is injected into the warm syrup, held at a temperature just above the melting point of gelatin (see Chap. 39 for details on the thermoreversible behavior of gelatin). In a marshmallow aerator, pins on a rotating cylinder (rotor) intermesh with stationary pins on the wall (stator) to provide the shear forces necessary to break the large injected air bubbles into numerous tiny bubbles that provide the smooth, fine-grained texture of marshmallow. The gelatin molecules in the syrup are preferentially attracted to the newly-formed air-water interface, where they adsorb and prevent the tiny bubbles from coming back together (coalescing). A continuous stream of light and fluffy marshmallow exits the aerator on its way to the forming step.

The marshmallow candy is typically formed in one of three ways. First, it can be extruded in the desired shape and cut into pieces, as done for Jet-Puffed marshmallows. Second, it can be deposited onto a belt, as done for Peeps. Finally, it can be deposited into a starch-based mold in a mogul (see Chap. 36) to make various shapes, including Santa and a Thanksgiving turkey (Bob the Builder was popular for a while). These are the chocolate-covered marshmallows popular at holiday times.

The temperature during forming is especially critical to get and retain the desired shape. Temperature needs to be just above the melting point of the gelatin so as soon as it's formed, it cools quickly and the gelatin sets, retaining the shape. If the marshmallow rope exiting the extruder is too warm, the marshmallow starts to flow before the gelatin sets. Instead of a round marshmallow, it will take a more oval form. In the same way, when the Peep shape is formed by depositing the marshmallow on a conveyor, the gelatin needs to set immediately or the result will be a flattened, shrunken Peep.

Because of the thermoreversible characteristic of gelatin, meaning it can melt and then reset, marshmallows are very sensitive to warm conditions. This year for our summer candy course, we had some marshmallow sent over from Europe (because they do more fun things with marshmallow there), but it was shipped in the summer without refrigeration. Rather than receiving a bagful of individual marshmallow pieces, they had melted and flowed together, and then reset once it had cooled. We had a bag filled with one single marshmallow.

That same thermoreversible behavior is important when you roast marshmallows to make s'mores. Getting it just right means you get a nicely browned marshmallow that hasn't fallen off your stick.

The modern marshmallow no longer bears any relation to the plant from which it's named. Nor does it have any particular health benefits. The sweet airiness of the marshmallow does, however, bring joy, or psychological health, to those who eat it.

51

Nougat

Probably one of the first sweets in history was a mixture of honey and nuts, blended together and perhaps dried to provide a relatively shelf stable confection. The first nougat was born.

Traditional nougat has been defined as "roasted seeds (almonds, hazelnuts, pistachios, pine nuts) kept together by a sweet paste made with honey, egg white, sugar and in some cases flavors." Although the specific origin of nougat is unclear, there is evidence of a nougat-like product during Roman times. Over the years, candy makers have played with the recipe to give us the nougat we know today. But the nougat you know depends on where you're from. Specifically, we'll compare European and American nougats.

Although nougat is widely available across Europe, two main focal points, some even say sources of origin, for nougat are Cremona, Italy and Montelimar, France. Cremona hosts a Torrone (Italian for nougat) Festival each year to recognize their spot in its history. Some say that it was invented for a medieval wedding in Cremona, with the name, Torrone, derived from the bell-tower shape of the candy piece served at the wedding. Montelimar also stakes a claim to the origin of nougat. In fact, nougat is sometimes called Montelimar, as noted in the Beatles song, Savoy Truffle (which was supposedly written to denote Eric Clapton's sweet tooth).

There are two main types of nougat found in Europe, crunchy and soft, depending on water content. The soft nougat is closest to what Americans think of as nougat. The United States version of nougat, not surprisingly, has been altered to meet American tastes and economic efficiencies. In fact, to many Europeans, what we

R.W. Hartel and AK. Hartel, *Candy Bites*, DOI 10.1007/978-1-4614-9383-9_51,
© Springer Science+Business Media New York 2014

consider to be nougat isn't nougat at all, even though it loosely fits the definition above.

The nougat used widely in American candy bars is typically sweetened with sugar and corn syrup, rather than with honey. It usually contains some vegetable fat to provide lubrication. And it often does not contain nuts. OK, when you break it down like that, maybe American nougat doesn't have very much in common with European nougat after all.

The old, discontinued Mars Bar in the United States was an exception, though, with a nougat looking much more like European nougat than any other American nougat. It contained an almond-laced white nougat, similar to European nougat, topped with caramel and coated in chocolate. However, the international Mars Bar, developed in England by Forrest Mars (son of Frank Mars), is much more like an American Milky Way bar. Further, the original American Mars Bar was discontinued in 2002 but brought back, sort of, as Snickers Almond. Confusing? You bet.

Two popular products that characterize the diversity of American nougat are Charleston Chews and the 3 Musketeers bar, known as chewy and grained nougat, respectively, based on the texture and eating characteristics.

Named for the dance popular at that time, the Charleston Chew was first introduced in 1922 and is now produced by Tootsie Roll Industries. As its name indicates, it is indeed chewy, the kind of candy that can "steal your teeth", as our old friend, Noda-san, an old-time candy maker from Japan, liked to say.

Chewy nougat is typically made by blending together a cooked sugar syrup with a whipped frappé to create an aerated product. A sugar and corn syrup mixture, heavily weighted towards the corn syrup to prevent sugar crystallization, is cooked to about 260 °F, to the firm ball state, with a water content of about 6–7 percent (see Chap. 8). It's put aside to cool a little while the frappé is made by whipping corn syrup with a protein stabilizer. Typically, frappé is made with egg whites or soy protein, or a mixture of the two. Whereas gelatin gives a spongy, elastic texture to marshmallow, egg albumin or soy protein gives a softer, more shaving cream-like texture to frappé.

The warm sugar syrup is then slowly poured into the frappé so as not to lose the air bubbles created during whipping. Once the syrup has been blended together with the frappé, the rest of the ingredients are added. In particular, a small amount of fat is added in most American nougats to help provide lubrication, both on processing equipment and on teeth. However, fat is known to reduce aeration, so the melted fat has to be carefully added to the batch at the end to make sure the candy retains the air bubbles. After the fat is added, nuts and other flavorings are gently blended into the thick candy mass until uniformly mixed. The batch is then poured out for forming and packaging.

Although there are air bubbles spread throughout the nougat mass, the sugar phase is amorphous and, in this case for sure, amorphous means chewy. When you pull a Charleston Chew apart, if you're careful, you can stretch it quite a way before the strand between the two halves actually breaks. That's chewy.

Compare that to the texture of a grained nougat, like the inside of a 3 Musketeer bar. Pull one apart between your hands and the strand breaks very quickly. Candy makers call that a "short" texture because the strand between the two pieces breaks off almost as you start pulling it. That's short.

By the way, the original 3 Musketeers bar was quite different from the one we know now. It had three individual pieces, or musketeers, in one wrap; that's where the name comes from. With vanilla, chocolate and strawberry, it was Neapolitan flavoring in one candy.

Making grained nougat is similar to the chewy nougat above, with a few key differences. For one, there is more sugar than corn syrup, in both the cooked sugar syrup and the frappé. And second, powdered sugar is the last thing added, to help seed crystallization of the batch. Both differences are designed to promote crystallization of the sugar to "shorten" the texture. Complete crystallization takes a day or so. The texture of a 3 Musketeer directly coming off the line is a lot different, a lot more like a chewy nougat, than the final product. Crystallization continues within the package until it reaches its final state in a couple of days.

When a candy bar with a nougat layer is manufactured, the nougat is formed into sheets through a set of calendering rollers. In a calender, a series of pressure rollers thins the layer of nougat down to the desired thickness for the particular candy bar. When necessary, another layer of candy, like caramel, would be calendered separately and then overlaid on the nougat layer. The multi-layered candy sheet would then go through a series of cutters in both directions to cut out strips of candy with the desired size, usually bar shape. Each of those bars would then pass through a chocolate waterfall to get enrobed, before entering a cooling tunnel to solidify the chocolate. At the exit of the cooling tunnel, individual bars are wrapped and cased.

Nougat is one of those candies where Americans differ from most of the rest of the world. You can find good European-style nougat here in the United States, but most Americans prefer the nougat in such products as Snickers, 3 Musketeers, Milky Way, Reese's Fast Break, Baby Ruth, Goo Goo Clusters, Pecan logs, and many more. On the other hand, if more Americans tried the traditional European nougat and sampled its delicate sweetness, perhaps it would be more widely appreciated and available.

52

Starburst

According to the Starburst Foundation, a nonprofit group with the aim of "preserving and protecting the ecosystem of our planet from natural or man-made disturbances," we're due for another burst of galactic superwaves to pass through the solar system sometime in the not too distant future (on a geologic timeframe). An intense cosmic ray particle barrage from the center of the Milky Way galaxy, the last superwave passed through around 12,000–16,000 years ago, initiating a span of mammal extinctions that was the worst since the dinosaurs flamed out. On the other hand, some scientists have theorized that these superwaves have led to spurts in human development through genetic mutations.

Starburst the candy, on the other hand, was first produced in the United Kingdom as a product called Opal Fruits. It wasn't until they were introduced in the United States in 1967 that the name was changed to Starburst because they're "unexplainable juicy™." How is a fruity, chewy candy like a starburst? Although we'll never know for sure, the name Starburst was probably the marketers attempt to say there was a burst of flavor in each bite. Besides, at that time, the space race was on full throttle and anything space was cool.

Fruit chews are lightly aerated products made with essentially the same ingredients as marshmallow, but in different proportions and with some added fat. There's sugar, corn syrup and water to start. The sugar syrup is heated to boil off some of the water and then cooled prior to addition of the stabilizer. Gelatin is often the stabilizer of choice, as in Starburst, but other stabilizers may also be found. Starburst contains some dextrin and modified starch that

R.W. Hartel and AK. Hartel, *Candy Bites*, DOI 10.1007/978-1-4614-9383-9_52,
© Springer Science+Business Media New York 2014

contribute to the texture of gelatin. A bit of fat is added in chewy candies to help provide lubrication in the mouth and make it easier to chew. Although it's definitely chewy, it doesn't stick to your teeth so badly because of the fat. The fat also helps us form and cut the candies, so the mass doesn't stick as badly to the machine surfaces.

Starburst candies, called a fruit chew in the candy industry, are slightly aerated to lighten the texture. Whereas marshmallows are highly aerated with a specific gravity of 0.3–0.5 (see Chap. 50), fruit chews are much denser. Drop a marshmallow into a cup of hot cocoa and it floats nicely on the surface. If you drop a Starburst into a cup of water it sinks (rapidly)—even though it's got air bubbles, it's not enough to allow it to float.

The gelatin helps stabilize air bubbles in Starburst, just as it does in marshmallow. As the candy mass is whipped to incorporate air, the gelatin molecules are drawn to the air interface where they protect the newly formed bubbles from collapsing. The gelatin also provides a unique chewy texture to the candy. Some versions of Starburst also contain starch and pectin to provide different chew characteristics.

Many chewy candies also contain some sugar crystals, called "grain" by candy makers, to moderate the chewiness and provide a slight shortness of texture. Pull the two ends of a Starburst apart and see where the strand between the two halves breaks. A highly grained confection like fudge has no strand; it breaks almost as soon as you pull because of all the sugar crystals disrupting the structure. A chewy candy like the caramel center of a Milk Dud can be pulled a long way before the strand breaks. Without crystals, there's nothing to break up the structure. Starburst is somewhere in between, but leaning towards the grained, short texture. The two halves break apart pretty quickly, indicating the presence of sugar crystals.

How is Starburst different from salt water taffy? No salt water, of course, although taffy doesn't really contain salt water as an ingredient either. There are also no crystals in taffy so it's stretchier than Starburst. And taffy is slightly more aerated, giving a lighter texture. Salt water taffy can also be found with different stabilizers.

Many versions are made with egg whites to provide a less elastic texture than derived with gelatin. Other taffy products use modified starch.

Starburst candy provides a really good example of how companies build on a successful brand with line extensions, defined as new products associated with a primary brand. For example, all the different versions of Cheerios are line extensions. The same goes for Starburst. They include Starburst Minis, Very Berry, Flavor Morph, FaveREDs, Original and Tropical Fruit Chews, Original Jellybeans, Crazy Bean Jellybeans, Sour Jellybeans, Tropical Jellybeans, and GummiBursts Original. There is even a version of Starburst candy corn. All carry the Starburst label, signifying that they're all part of the main brand and leading consumers who enjoy the original to try out these newest candies.

Starburst Minis are the newest line extension, essentially small bits of the original Starburst. Even though the original candies are pretty small, Minis are about one-third the size, and they come unwrapped. They also don't use gelatin as the stabilizer, they use pectin and so would be a little less chewy than the original. Because they're unwrapped, they're also more prone to changes in moisture. Without a wrapper, on a warm, humid summer day, they'd pick up moisture and turn into a sticky mess. On a cool, dry day, they'd start to dry out and get firm, eventually turning into nuggets of hard rock chewies (would that be another line extension, specially designed for those with good teeth?).

Playing on some recent flavor encapsulation technology, Starburst FlavorMorph, a relatively new addition to the Starburst family, starts out with one flavor and changes to another. The first flavor is located in the main bulk of the candy, but then as you chew it, you break through the beads encapsulating the second flavor, causing the taste to change. Flavor changes are orange to orange strawberry and cherry to cherry lime. Note that the original flavor is still there as the new one is released. You can only do so much with that technology.

What will the next Starburst line extension be? Only the product developers know for sure, but it most likely will have nothing to do with the Starburst Foundation or the galactic superwaves that are predicted to affect the human race in the near future.

53

A Whopper of a Story: Malted Milk Balls

Many confections started out with a medical or health background. So the story goes for malted milk balls. Well, sort of anyway. It's the primary ingredient that distinguishes malted milk balls—malted milk—that has somewhat of a health background.

Malted milk was developed in Racine, WI in 1887 by a British pharmacist, William Horlick, looking for a healthy infant food supplement. It was a natural in Wisconsin, where brewing and dairy are two of the main industries. A by-product of the brewing industry, malt extract, was added to milk to create a uniquely flavored healthy and natural product. Horlick chose infant nutrition as the health target.

Making barley malt involves three main steps—steeping, germinating, and drying. The barley grains are first steeped in water until they reach about 45 percent moisture. The wet grains are then allowed to germinate, or sprout, for up to seven days, developing a unique enzyme system in the process. The germinated grains are then dried at about 130 °F in a kiln (or oast, for you crossword fans) under conditions that retain the enzyme activity. These enzymes are critical for developing sweetness and flavor.

To make malt extract from barley malt, the mash process is used to extract the sugars. When the malt is mixed with warm water, the enzymes, primarily the amylases, are activated and begin to cleave the longer starch molecules (amylose and amylopectin) into smaller sugars. The process is very similar to enzyme production of corn syrup from corn starch (see Chap. 12). The aim of mashing is to create a syrup with about 10 percent glucose, 40 percent maltose (two glucose molecules joined together), and the rest longer-chain

R.W. Hartel and AK. Hartel, *Candy Bites*, DOI 10.1007/978-1-4614-9383-9_53,
© Springer Science+Business Media New York 2014

saccharides (longer polymers of glucose). At this point, the product is heated to destroy the enzymes and prevent further changes.

The thick liquid product from mashing is called malt extract. Because of its saccharide profile, it's about 50–60 percent as sweet as sucrose. Due to its similarities to corn syrup, malt extract is gaining popularity as a natural sweetener in various industries, including confectionery. It can be used either as a thick liquid or in dried form.

Malted milk powder is made by mixing wheat flour and malted barley extracts, whole milk, sugar, salt, and baking soda. Essentially it's a powdered drink mix, a lot like Ovaltine. With that characteristic unique malty taste, Ovaltine makes good use of malt extract. It was developed in 1904 in Switzerland, a few years after Horlick's malted milk powder became available.

There are several methods to make malted milk powder. One version is simply to mix the dried powders (dried malt extract, dried whole milk, sugar powder, baking soda) until well blended. Another method is to dissolve the malt extract, sugars and baking soda in the milk before concentrating in an evaporator to make a thick slurry. This slurry is then dried in a vacuum dryer, with the powder being pulverized, classified, and packaged.

Malted milk powder can be used in a variety of products, although the most common are malted milk shakes and malted milk balls, where it provides a unique flavor derived from the malting process. Although the use of malted milk powder in confections probably began much earlier, malted balls were first marketed as Giants in 1939 by the Overland Candy Company. The company was taken over by Leaf, who renamed them Whoppers. Hershey now makes Whoppers, probably the most widely-known example of malted milk balls.

Malted milk balls are an interesting product, not just because of their unique malted flavor. The manufacturing process is itself unique, another example where technologists were able to produce a product by experimentation, without really understanding the science behind what they were doing. The development process also included a step that was not intuitive, one that isn't common at all in confections. Perhaps the developer came from another industry and asked himself what would happen if.... Many so-called "new" discoveries are actually just a translation of a technology or idea from one field to another.

Malted milk balls comprise three layers—the malted milk ball center, the chocolate coating shell, and the polish layer. The malt ball center is where it all starts. The candy maker mixes corn syrup (although malt extract would work and give an extra malty flavor), sugar, malted milk powder, and a few other ingredients, and then cooks it to the right temperature to get a sugar mass with about 4–6 percent water content. This is a bit softer and more pliable than a hard candy mass, since it's only cooked to the soft crack stage (see Chap. 8). There's also a little fat used here to help lubricate the mass during processing.

Once the mass has cooled to the consistency of moldable plastic, it's aerated on a candy pulling machine, the same type used for taffy. Three inter-rotating arms grab, pull, and fold the candy over and over on itself, creating air pockets dispersed throughout the plastic mass. Once sufficient air has been incorporated, small balls are cut from the mass by passing it through a drop roller. Imagine two metal cylinders with a series of indentations, counter-rotating so that the indentations line up during rotation. The still plastic mass is forced between the two rollers, which create candy balls, or drops, connected by a candy webbing. The candy is quickly cooled to create semi-solid, aerated candy balls and the webbing recycled to the next batch.

The next step is where the creativity came in. What caused the candy maker to think about putting these balls into a vacuum oven is unknown, maybe the same curiosity of a kid with a marshmallow Peep and a microwave oven? In the oven, the candy warms up and becomes more fluid-like while the vacuum causes the air cells to expand, a la the ideal gas law. The combination causes the candy balls to puff up, increasing in size by two- or threefold (that same principle is used in Peeps jousting by the way). The next key step is that the temperature must be reduced while the vacuum is still maintained to allow the candy syrup mass to set up into the glassy state around the air cells. The result is a crunchy, malted milk flavored candy delight. It's essentially a glassy sugar candy matrix holding in the air bubbles caused by the vacuum.

The chocolate coating layer is applied in a traditional rotating pan by drizzling melted chocolate over the tumbling pieces. The tumbling action spreads the melted chocolate to uniformly coat each piece and allow it to set into a solid chocolate coating. The piece is finished by polishing it with wax and/or shellac to give it a nice shine. In this way, it develops an appealing appearance to go along with the malty goodness.

While malted milk balls may not be the healthiest product, malt extract notwithstanding, they're certainly a tasty treat with a unique flavor from the land where cows and beer come together.

54

Retro Candy: Bit-O-Honey and Mary Jane

Retro candy is popular these days, at least some of them. Candies that were a big hit decades, even a century, ago often have a nostalgic appeal. And some of them are still being made. Two such candies are Bit-O-Honey and Mary Jane, both taffy-like candies with somewhat unique flavors.

The Mary Jane is the eldest of the two, first developed in 1914 by the Charles Miller Company in Boston. As one claim to fame, the Miller Company got its start in 1884 selling candy out of the same house in which Paul Revere lived until 1800. Named for his favorite aunt, Miller combined molasses with peanut butter to give the Mary Jane a unique and unusual flavor. Molasses used to be a quite popular flavoring and sweetener, although now its use is limited to very specialized occasions, and only rarely in candy. Today, Mary Janes is part of the NECCO brand of candies.

Bit-O-Honey was developed in 1924 in Chicago by the Schutter-Johnson Company. While not as colorfully named as the Mary Jane, Bit-O-Honey has a similarly chewy texture that's flavored with almonds and honey. The rights to Bit-O-Honey have been bought and sold numerous times (see Chap. 4). It was purchased by Nestle in 1984, who just recently sold the brand to Pearson's Candy Company of Saint Paul, Minnesota.

It's interesting to compare these two retro candies side by side, as we often do in our candy class. Looking at each of the bite-sized candies, one sees some similarities. First, they're both about the same size, rectangular-shaped candies wrapped in a plastic wrapper. In the Mary Jane, the wrapper is carefully folded over, reminiscent of how we wrap Christmas packages, while Bit-O-Honey has twist

R.W. Hartel and AK. Hartel, *Candy Bites*, DOI 10.1007/978-1-4614-9383-9_54,
© Springer Science+Business Media New York 2014

wraps on each end, somewhat like the Tootsie Roll midgees. Unfortunately, neither packaging approach provides good water barrier protection, so both candies tend to dry out quickly over time. What starts out as a tender chewy piece, similar to a fresh salt water taffy, can end up as hard as a Sugar Daddy caramel. Working on an old, dried out Bit-O-Honey (or Mary Jane for that matter) takes the same patience as eating a Sugar Daddy. I can only imagine how tasty a fresh one would be right off the line. Since flavor release is enhanced with a softer matrix, my guess is the delicate honey flavor comes through so much better in a fresh candy.

Both are wrapped in plastic layers that are bright yellow with red stripes. Mary Jane has its red stripe around the mid-section of the candy while Bit-O-Honey has red bands running along the length of the piece. In order to see how bright the yellow color is, you'll have to remove the candy from the wrapper. In both candies, the brown color of the candy comes through the wrapper, giving the wrapped candy more of a muted yellow appearance.

When removed from the wrapper, both candies are a tan brown color, one from the peanuts and the other from almonds. Mary Jane candies are a little darker, perhaps due to the roasting of the peanuts, and they're also a little larger. The size difference is hard to compare directly though, in part, due to a difference in density.

Both are slightly aerated candies, similar to chewy nougat candies (see Chap. 51), but both are much denser than the typical salt water taffy. In the same way as nougat, Bit-O-Honey contains nonfat milk, egg whites and modified soy protein to help stabilize the air bubbles incorporated to reduce density. The Mary Jane contains no proteins, relying on the high viscosity of the matrix to hold air bubbles. Based on texture and weight, the Mary Jane seems to be a little denser than the Bit-O-Honey.

Pull each candy apart and you see considerable difference in how they stretch. The Mary Jane stretches and stretches and stretches before the strand between the two pieces finally gives way, unlike the Bit-O-Honey, where the strand breaks after just a slight pull. This "short" texture is characteristic of a slightly grained (or crystallized) matrix, similar to that of the Tootsie Roll (see

Chap. 31). This suggests that Bit-O-Honey also contains a small amount of sugar crystals to help break the strand as the two ends are pulled apart. Alternatively, finely ground almond pieces could impart the same type of short texture.

These differences explain the contrast in texture when the two are chewed. Both are chewy for sure, enough to pull out fillings, but the Mary Jane is definitely worse for sensitive teeth. The Bit-O-Honey is so much easier to chew and sticks less to your teeth. The proteins that hold air bubbles and the particles to break up the stretchiness provide the Bit-O-Honey with that easier chew. So, at first glance, Mary Janes and Bit-O-Honey may seem very much alike, but in the end their textures are considerably different.

Of course the flavors are different as well. Almonds and honey for Bit-O-Honey versus peanuts and molasses for Mary Jane. Both are classic favorites with age-old appeal.

A couple other candies that fit the retro chewy candy category but that may not be as well known are Squirrel Nut Zippers and peanut butter kisses. If you're from the Boston area, maybe you've heard of Squirrel Nut Zippers, but almost everyone knows peanut butter kisses. You either hate them or love them.

Squirrel Nut Zippers, originally made by the Squirrel Brand Company in Cambridge, Massachusetts, first appeared in 1926. The origin of the name is pretty interesting, even if only partially true. The story goes that the Squirrel Brand owners read in the newspaper that a drunk arrested during Prohibition blamed his intoxicated condition on a Nut Zipper, a drink popular in Boston at that time. They liked the name so much that they called their newly developed candy the Squirrel Nut Zipper. A music group in the 1990s also liked the name so much that they chose Squirrel Nut Zippers for their band name. They even gave out the candy at concerts.

Squirrel Nut Zippers, name notwithstanding, are actually a peanut butter-flavored caramel. Except for the addition of peanuts, the ingredient list reads just like a caramel—corn syrup, sugar, peanuts, condensed milk, fat, and so on (see Chap. 28).

Peanut butter kisses, those taffy-like candies wrapped in orange and black twist wrappers, are mostly popular at Halloween. Although Wisconsin's Melster Candies makes them, one of the major producers of peanut butter kisses is NECCO under the Mary Jane brand. Since they're essentially molasses-flavored salt water taffy with a ribbon of peanut butter through the center, they fit right in with the Mary Jane brand.

What keeps these candies hanging on while others end up on the dead candy list (see Chap. 64)? It's hard to say, but having a fan base who come back over and over again is a necessity; since candy making is a business, a company will continue to produce a brand only as long as they make a profit. Maybe as the older crowd disappears, so will the Bit-O-Honey and Mary Jane. So if you want them to stay around, go out and buy some, and bring some kids along.

55

Gum Wads

Ever notice all those black spots on the sidewalk, especially in front of public buildings? Most of those are gum wads, discarded after all the sweetness and flavor has been chewed out of them. Some gum chewers, like some cigarette smokers, like some people in general (especially in movie theaters and stadiums), seem to think it's OK to throw their waste on the ground for someone else to deal with.

Look at all those white, pink and black spots, gum wads, on the sidewalk from people just tossing them out (like cigarette butts in the gutter). Fresh wads start out the color of the original gum, but over time they turn black as the chemicals in the gum base oxidize when exposed to light, leaving an unsightly speckled sidewalk.

Discarded gum wads got to be such a problem that Singapore actually banned chewing gum in 1992. Story has it that chewing gum stuck on the doors of a train car actually kept the train from running. This led to a ban on the import, sale, chewing and even possession of gum. That ban is now being lifted somewhat, but only for people who need to chew gum for medical reasons and is almost as highly controlled as prescription drugs.

Banning gum entirely may be a little extreme, but it shows how serious the problem of gum disposal can be. It's enough of a problem that gum makers have taken notice and have started working on solutions, without curbing that crave to chew of course.

The reasons for chewing gum are many—for the sweetness and flavor, to reduce stress and maybe even to quit smoking. But, after about ten minutes of chewing, we've extracted all the sweeteners and flavors that we're going to get out. We're left with a wad of gum base—the cud or bolus. The gum base, primarily made up of

R.W. Hartel and AK. Hartel, *Candy Bites*, DOI 10.1007/978-1-4614-9383-9_55, © Springer Science+Business Media New York 2014

rubber, is the ingredient that makes gum chewy and the reason the wad sticks to the pavement.

Supposedly, General Santa Ana, of the Alamo fame, brought some natural chicle (rubber from the chicle tree) to New York, and that led to the development of chewing gum as we know it. However, the original chicle as a source for gum base is expensive and, as with most agricultural products, highly variable. So the majority of today's gum contains synthetic rubbers, or elastomers, instead of chicle. Interestingly, rubber bands are made from the natural chicle because of its desirable elastic properties. In gum, the synthetic rubbers, in combination with various additives, are designed to mimic the physical properties, like chewiness, softening and melting, of the natural chicle.

Gum base is a regulated substance, sort of. It has a Standard of Identity, with specific composition as defined in the Code of Federal Regulations. Typically, it contains rubbers plus numerous additives, such as texture modifiers, plasticizers, softeners, fillers, emulsifiers, and antioxidants, to give specific properties to the gum (like chewing versus bubble gum). In fact, the last item on the list of approved components of gum base includes any "substances generally recognized as safe." You can add almost anything to gum base as long as it's approved for human consumption. Not really very regulated at all, despite the Standard of Identity.

The main components of gum base are the rubbers, particularly butyl rubber, styrene-butadiene rubber, and polyisobutylene. Lower molecular weight polymers, including synthetic polymers such as polyvinyl acetate and vinyl acetate-vinyl laurate copolymers, are also typically added to gum base to impart specific properties. While chewing gum is not the same as chewing car tires, they do actually use some of the same rubbers.

In fact, gum base is one of the most tightly guarded secrets in probably all the world. Getting specific access to what's in any company's gum base is impossible unless you're on the inside. And probably even then it's only on a need to know basis. The gum business is probably even more competitive and highly secretive than the chocolate business.

What's the difference between chewing and bubble gum? The ingredient lists for two comparable gums are almost identical, but you can hardly blow a bubble with chewing gum. What's bubble gum's secret?

It's all in the elastomers, the nature of the rubber compounds. Bubble gum contains a slightly different type of rubber polymer than chewing gum. There are more, longer (higher molecular weight) rubber molecules in bubble gum than those in chewing gum. The longer polymers stretch more easily and hold the bubble better.

It's those nonbiodegradable rubbers in gum that cause a discarded wad to stick and stay, seemingly forever, on the sidewalk. They're nearly impossible to remove. In fact, it takes a jet of high-pressure steam to get rid of those sticky wads. Because of this, most gum manufacturers have been working on biodegradable and non-stick gums for many years. A few such products are already on the market and it's certain we'll see more over the next few years. This drive has also led to a resurgence in the use of natural chicle in gum since it is inherently more biodegradable.

Instead of spitting it on the sidewalk, people sometimes swallow their gum wad. Does it really stay in your stomach for seven years?

Nah, usually, gum passes right through you just like any other indigestible substance. It usually takes a couple days to be broken down and clear the system. However, swallowed gum has occasionally caused medical problems. There are reported cases of esophageal or colonic bezoar (a medical term for something that gets hung up in our digestive system, sort of like a hairball) where a gum wad got stuck and prevented normal digestion. One woman had been swallowing five wads of gum a day for a few years and had built up one of those bezoars so that it had to be surgically removed.

If you shouldn't spit gum out on the ground or swallow it, I guess the only thing left to do is put it under the seat, right? No, that's gross. Wrap it in paper and throw it away, please.

56

Gumballs

The gumball machine is an icon at the exit of convenience stores and other retail outlets. It provides an interactive event between a kid and his gumball.

Spitting out a ball of sweet chewing pleasure for a penny, or now, a quarter with inflation, these were one of the earliest vending machines, at least for confections (see Chap. 59). In fact, gum dispensing machines were around as early as 1888, although those gave out stick gum. The first to dispense a gumball was found in 1907.

Most gumball machines are similar—a glass dome filled with shiny sweet orbs of either chewing or bubble gum with a mechanism that trades a coin for the treat. In older machines, a coin is placed in the slot and the entire mechanism rotated to deposit the coin in the bank and release a gumball to a gravity-fed slot. But, as we'll see, gumball machines have advanced significantly over the years.

Gumballs may magically drop out of a machine into a waiting kids hand, but a lot of work went into making that chewy sphere. Compared to making a stick of gum, making a gumball takes a lot more care and effort.

All gum is essentially a mixture of gum base, sweeteners, colors, flavors, and a few other ingredients that enhance the chewing pleasure. Gum base, a complex mixture of rubber polymers and texture modifiers, is the starting point (see Chap. 55). To make gum, each of the ingredients is added sequentially to warm gum base until it's all uniformly mixed. Finely ground sugar powder and a little corn syrup are added to provide sweetness and structure.

R.W. Hartel and AK. Hartel, *Candy Bites*, DOI 10.1007/978-1-4614-9383-9_56,
© Springer Science+Business Media New York 2014

Colors and flavors are blended in to provide chewing enjoyment, perhaps with a little glycerol as well to keep the gum soft. Organic acids may be added here too, to provide tartness and to brighten up the flavor. Depending on the type of gum, a high intensity sweetener may be added to enhance and extend sweetness and flavor.

Mixing all the ingredients into gum base takes a huge motor with tons of torque to make sure everything is mixed well. In the lab here we have what's called a sigma blade mixer for making gum. The working chamber where the mixing takes place is almost dwarfed by the rest of the unit, primarily the huge motor that turns those blades and the framework to hold up that motor.

The design of the blades is critical to getting adequate mixing. The two blades, formed roughly in a z-shape (or sigma shape), are counter-rotating and mesh together as they turn to force the mass down in the middle between the blades. The mass comes back up to the top along the edges, where it's moved back towards the middle and again down between the blades. The design is efficient for mixing rubbery and viscous material, like gum. It's cool to watch it mix bubble gum—because of the way it folds and moves the mass, it actually blows, and pops, bubbles while mixing.

Once all the ingredients are mixed together, the gum has to be formed into the desired shape. An extruder forces the warm mass through an orifice, or die, shaped to give the right form. For example, stick gum is extruded through a slit die into a sheet, which is then scored and cut. To make gumballs, the gum mass is extruded, often vertically downwards, through an annular (in the shape of a ring) die to form a hollow tube or rope of gum. Air is blown into the middle of the rope to keep it from collapsing.

The warmed gum mass is semi-fluid at the point when it exits the die and immediately starts setting as it cools down. It's carefully collected on a conveyor and moved along to the cutting machine where the balls are created. Temperature is crucial since the gum must still be sufficiently pliable to cut and form, yet solid enough that the shape doesn't deform after the ball is cut. A section of the tube is cut to a specified length and, one at a time, each section is allowed to enter the ball cutter. A ball cutter contains three series of rotating blades. Two of the blades rotate in one direction to cut a length of the tube and the third set rotates perpendicular to the other two to smear the edges and form a ball. You almost have to see it to believe it—in goes a hollow tube of gum and out roll a bunch of gumballs.

The gumballs drop from the cutter onto a series of cold tables or conveyors where they're kept bouncing slightly to allow them to cool without creating a flat spot. Once cooled to the point where they no longer deform, the gumballs are then moved into a storage area to set completely prior to application of the sugar shell in the panning room (see Chap. 45). After they've been polished to give the nice shiny appearance, the gumballs are ready to be loaded into the gumball machine.

Gumball machines come in all sizes and many different types. You can buy a small toy version for home, without the need to pay for each piece. Or you could buy a colossal one—the jumbo gumball machine is nearly seven feet tall and holds 23,000 gumballs. If you had one of these, you'd always be the hit of the party.

Interactive gumball machines are available that provide an even greater experience. Some units have complex spirals or chutes, sometimes with flashing lights, to deliver your treat from the dome to the pick up chute. But the really interactive ones allow you to actually use the gumball before you chew it, through either a sports event or as a pinball machine. Drop your quarters into the slot and a gumball drops into the pinball game chute. You use that gumball to play pinball until you lose and the gumball drops out for you to eat. You can even win additional gumballs if you're a pinball wizard, good with the flippers. You can't lose.

57

Gum and the Bedpost

Does chewing gum lose its flavor on the bedpost overnight? While it seems a simple question, it's actually a pretty complex issue. Flavor is one of the main reasons we chew gum, but the relationship between flavor release and gum chewing is complicated. Add in the bedpost, and it's even more complicated.

First, you might ask, is there really a reason to put your partially-chewed gum on the bed post overnight? Or behind your ear for later? Or on the side of your plate for after dinner? I don't know, maybe; there are numerous reasons why people chew gum and saving it for later seems like a prudent thing to do. Especially if there is some flavor left.

Besides the sweetness, the flavor keeps people coming back to their favorite brand or trying the never-ending stream of new products constantly being introduced by gum makers to attract our gum-chewing dollars. Gum makers are also continually looking for ways to extend the time over which flavor is released—numerous "long-lasting flavor" products have been emerging on the market as new technologies for protecting and releasing flavors are developed.

When gum is made, flavors are added in the mixing step and incorporated uniformly throughout the gum matrix. But that matrix is dominated by the gum base and its characteristics (see Chap. 55), particularly how it interacts with the flavor molecules. Flavor molecules are typically small and very volatile, meaning they quickly evaporate into the air—we sense them in our nasal passage so technically we should be calling them aroma molecules (although we'll continue to use the more traditional term here).

R.W. Hartel and AK. Hartel, *Candy Bites*, DOI 10.1007/978-1-4614-9383-9_57,
© Springer Science+Business Media New York 2014

Flavor molecules can either be hydrophobic (don't like water) or hydrophilic (like water). Since gum base itself is hydrophobic, the flavor molecules that don't like water prefer to reside in the gum base—a lot, to the point where it's actually pretty tough to get them out.

When you put a chunk of gum in our mouth and start to chew, lots of complex things start to happen. Chewing gum, for being such a simple thing, has an amazing amount of science behind it, especially in flavor release.

When we chew gum, or any food for that matter, there is a complex sequence of events that occur between the gum and saliva, between the gum and the air, and between our saliva and the air. These events are governed by both thermodynamic principles, primarily how the flavor molecules partition between gum, saliva and air, and the kinetic rates that the flavors pass from one place to another. To create a gum with longer-lasting flavor requires an understanding of both.

It all starts with the sweetener, in the form of small crystalline particles distributed uniformly throughout the gum matrix. As you start to chew a piece of gum, your saliva begins to hydrate the gum, causing some of the sweetener to dissolve. The sweetened saliva then interacts with the appropriate taste buds on your tongue to provide the sweet sensation (see Chap. 7). At the same time, some of the flavor molecules release into the saliva and then into the air, or directly into the air, to be snorted up into your nose to provide the aroma. Both hydrophobic and hydrophilic flavors are released, but especially the water-loving ones pass through the saliva, an aqueous solution of proteins (namely, mucin). As long as there's sweetener to be dissolved, there's flavor being released through these processes.

Once the sweetener has all been dissolved, gum scientists confirm that there is actually still plenty of flavor left in the cud (the wad left in your mouth). When chewed gum has been broken down and chemically analyzed for what flavors remain, typically only about half of what was initially added to the gum has been released. Up to half of the flavor still remains in the cud! But you can't access that flavor because, thermodynamically, it prefers to remain within the hydrophobic gum base. Emulsifiers used in gum help hydrate the gum base and help with flavor release, but only up to a point.

When you spit that cud out onto the ground, put it under your chair or, better yet, wrap it in paper and throw it away, you're actually tossing out some of the most expensive components of the gum—the flavors. It's a waste, and that's why gum makers are working hard at trying to deliver more flavor for a longer time. And why it might be worth putting it on your bedpost for later.

Gum makers have two approaches to make longer-lasting flavor—they can either protect the flavor so it releases more slowly or they can protect the sweetener for the same reason. Usually they do both. Encapsulation technology is the key. Both flavors and sweeteners can be encapsulated to better control their release. It's similar technology to controlled-release medical applications, just refined for use in gum.

There are numerous approaches to encapsulating flavors. One of the most common is simply to spray dry the flavor with a high molecular weight carrier like maltodextrin (long starch-based molecules). The flavor is dispersed within the solidified matrix of the powder particles, only to be released when the maltodextrin dissolves, slowly in your saliva as you chew the gum. Flavors can also be encapsulated in protein-based beads. The flavor may be retained inside a tough gelatin shell, to be released only in the warmth of the mouth, similar to a gel-cap medicine. Sweeteners are also encapsulated to provide added flexibility to control flavor release.

Gum makers have numerous options to extend gum flavor release. However, flavor technology is one of the most highly secretive facets of modern gum manufacture, right up there with the secrets of gum base. Everything is patented and secrets are kept under lock and key.

But now let's look back at that gum wad stuck on your bedpost. You scrape it off the next morning, stick it in your mouth and chew. Does it have less flavor than the night before? Almost certainly it will have a little less of the original flavor, but it will undoubtedly have more bedpost flavor too. After it's been chewed, the hydrated gum wad you put on the post was free to equilibrate with the air. Since there were no gum flavor molecules in the air around your bed, some of those in the gum will be driven to reach the air. With the extra water from chewing, the rate of diffusion out of the gum will also be enhanced, but only slightly because most of the flavor molecules still prefer to stay in the gum base. It might also have dried out a little overnight as it lost a little moisture to the air.

The end result? Well, it depends. If you had already chewed out all the flavor that was extractable, leaving it on the bedpost won't bring it back. If there was still extractable flavor left, much of that would still be available once you started chewing it again.

58

Medicinal Gum

Chewing gum is a good way to release stress. Ball players relieve the stress of anxiety in the game by chewing Big League Chew, a gum developed specifically by big league ball players who found chewing tobacco to be offensive. The mindless smack, crack and pop of gum chewing, similar in a sense to mindless leg jiggling, provides a physical outlet when other ways of relieving stress (like pacing back and forth) are not feasible.

But, for others, listening to people chew their gum noisily increases their own stress level. Ever notice how some people chew gum with their mouth open, like cows chewing a cud? Or snap their gum loudly. Or pop bubbles loudly and indiscriminately. Usually without even thinking about it. Especially without thinking about how the people around them might be affected.

There are medical reasons to chew gum, including to freshen breath, whiten teeth, reduce xerostomia (dry mouth), and to release some active compound (as in Nicorette). Some airline passengers chew gum to relieve pressure on the ear drums, particularly during take-offs and landings. Some chew for the simple fun and enjoyment, and, of course, it's a sweet treat.

Gum makers understand all of these reasons; their job is to find ways to satisfy them. The scientists at Wrigley and other gum companies study the complex interactions in gum in order to better meet our needs. Newer and better methods of delivering a satisfying gum chewing experience, including delivery of active ingredients, are found as technology continues to develop.

Almost right from the start, gum was used to promote health. Beeman gum, first sold in 1879, contained pepsin. Pepsin is an

R.W. Hartel and AK. Hartel, *Candy Bites*, DOI 10.1007/978-1-4614-9383-9_58,
© Springer Science+Business Media New York 2014

enzyme normally present in our digestive system that specifically breaks down proteins into smaller peptides. Thus, it's touted as a digestive aid, which was how Beeman gum was marketed. There probably wasn't a lot of science engineered into Beeman gum; it was essentially chicle and pepsin powder, probably not even sweetened. However, the marketing of Beeman was classic, and may be the start of the obesity problem. On the package was an image of a pig, with the slogan "With pepsin, you can eat like a pig." Boy, things have changed in the past hundred or more years. Can you imagine someone coming out with a slogan like that, seriously?

Another reason people chew gum is to whiten their teeth. Gum makers use carbamide peroxide to provide the whitening effect, but since it reacts with other components of gum (water and high intensity sweeteners), it needs to be protected to retain activity for the entire shelf life of the gum. In the presence of water, as in contact with saliva, carbamide peroxide breaks down to produce a peroxide radical, which provides the bleaching effect (just like when people use peroxide to bleach hair). Since carbamide peroxide is water sensitive, it is best used in gel-based products like toothpaste. To use it in chewing gum, carbamide peroxide must be protected from the water in the gum itself.

In a similar technology to controlling flavor in gum (see Chap. 57), protecting carbamide peroxide from water in gum requires encapsulation technology. Solid crystalline particles of carbamide peroxide are encapsulated in some sort of wall material. In fact, one recent patent on the subject suggests that two thin layers of encapsulant material were most desired.

The first chewing gums were sweetened by sugar, but it was evident that chewing gum contributed to dental problems. Despite the benefits of chewing gum, sugar reduces the pH in the mouth and promotes microbial growth, which result in loss of enamel and tooth decay. As alternative sweeteners were being developed in the 1950s and following decades, their use in gum was inevitable. As it turns out, the sugar alcohols (see Chap. 18) that make up the bulk sweetener in sugar-free gums help protect against tooth decay—a great marketing tool for gum makers. Common sugar-free gum ingredients like sorbitol and maltitol are noncariogenic (do not promote tooth decay) while one, xylitol, has actually been proven to be anti-cariogenic (fights tooth decay) and can be declared as such on the gum packet. As the old Trident ad went, "It's the only gum my mom let's me chew." For good reason.

Although sugar-free gum has documented advantages over sugar-based gum, it has its down side as well. Chew too much of it and you'll find yourself running continually to the bathroom. The sugar alcohols are nondigestible, which is why they have less calories per gram than sugar. But as they go right through you, they carry water with them, leading to diarrhea and intestinal problems. Another example of too much of a good thing is bad for you.

Another target for some gum products is salivation—increasing saliva flow to head off cottonmouth. Xerostomia, or dry mouth (or pasties, drooth or doughmouth), is a medical condition often related to reduced salivary flow. A normal person generates a flow of 0.3–0.4 milliliters (mL) saliva per minute, with below 0.1 mL per minute being abnormally low. One gum, Quench, was developed in the 1970s by a former Wisconsin athlete specifically to fight cottonmouth associated with sports. The main ingredient for salivation is nothing more than citric acid, a lot of it. Citric acid is the third ingredient, behind gum base and sugar. The high acid content requires careful design of the gum base to prevent undesirable reactions over time and makes it quite tart, of course.

As we've shown, there are numerous reasons for chewing gum. If that's not enough, Stride gum brings us another reason—a video game. Seriously, you can download an app for certain phones called Gumulon where the mobility of Ace, the main character, is controlled by your chewing action. The phone videocamera keeps tabs on your chewing rate, converting that into Ace's movements. The aim is to keep Ace from falling into the jaws of the monster—getting chewed up as it were.

But remember, as you're saving Ace in Gumulon, or just chewing like you usually do, please be sensitive to how your chewing behavior affects those around you.

59

The Vending Machine*

It's two o'clock on a Thursday afternoon. You can feel your eyelids slipping; the words on the page start to run together. Your stomach rumbles, begging to be fed. So you grab a handful of change and head off to the vending machine for a sweet afternoon pick me up.

While some people can pony up to the vending machine and instantly pick a tasty treat, I'll spend the next five minutes staring through that glass window thinking, "Now what do I want? Milky Way? No, no, too sweet. How about Twix? Oh, they have Snickers! Reese's Peanut Butter Cups! What do I want?" After five minutes of deliberating, I put money in the slot and punch in the numbers for the perfect snack to survive the afternoon.

Customers of the first vending machines didn't have so many choices. Built in the first century, the machine dispensed holy water when a coin was inserted. The machine was designed to stop people from taking more than they had paid for. The amount of holy water dispensed depended on how heavy the coin was. While the first vending machine was a cool trick, it didn't exactly take the world by storm.

Modern vending machines were invented in the late 1800s. They sold post cards and gum. Some machines added mechanized figures that would perform a little show whenever a coin was inserted, technology that led to the development of slot and pinball machines. Soda and cigarette machines weren't far behind.

The ultimate example of choice in vending machine food was the automats of the early 1900s. An automat was a self-service

---

*Primarily by AnnaKate Hartel

R.W. Hartel and AK. Hartel, *Candy Bites*, DOI 10.1007/978-1-4614-9383-9_59,
© Springer Science+Business Media New York 2014

restaurant where diners would insert the right amount of money, slide open a small glass door, and grab their food. Though introduced in Philadelphia, they had their heyday in post war New York. But the introduction of fast food caused the automats to slowly close until they disappeared. However, variations have been popping up in Japan and the Netherlands. Perhaps the automat is due for resurgence in the United States.

Even Thomas Edison tried to get into the vending machine game. He imagined a completely coin operated general store. Insert the correct coinage and out would pop anything from a bag of nails to a yard of lace. Unfortunately the mechanism never worked and he never built a prototype.

Perhaps Edison was just a little ahead of his time. Today, it's possible to get just about anything from a vending machine, just like he envisioned. Public restrooms around the United States stock personal items while malls are increasingly adding vending machines selling everything from face wash to electronics. While cigarette and newspaper machines are increasingly rare, they once were important fixtures in the American landscape. Machines that rent movies and games are in grocery stores and pharmacies.

But compared to other countries, the United States is pretty tame in terms of vending machine varieties. If I were standing in front of a vending machine in, say, Australia, I could be choosing which gem stone I wanted. Machines in Sweden can dispense library books, which helps promote literacy in rural areas. In many major cities bikes can be hired by the hour. Machines like this were just installed on the crowded streets of New York City. In countries where a biking culture is already established, like the Netherlands, bike parts are also available by vending machines.

But the vending machine capital of the world has to be Japan. With limited space and a very high population density, there is about one machine for every 23 people—Japan has the most vending machines per capita. While most machines sell drinks and snacks, some have more exotic fare. Eggs, bacon, and potted plants have all been seen in vending machines around Japan. There's even a claw machine that dispenses lobsters, if you're skilled enough to catch one. Catching a stuffed animal is already pretty tough, imagine trying to hook one that has its own claws.

Japan is also the leader in making vending machines more selective in order to sell adult items like cigarettes and alcohol. Smart cards identify a person as being of age and must be used when making purchases from those machines. This ensures that no minors have access to contraband items. But these cards aren't foolproof and some substances require even stricter security measures. Machines in southern California use fingerprinting technology to sell medical marijuana, so only the patient has access to this machines controversial product.

Vending machines have gone a long way from just dispensing a measured amount of holy water and the improvements keep coming. Today's vending machines customers are clamoring for the freshest food possible and inventors are working to accommodate that. While the ability to make fresh products isn't new—coffee machines have been brewing fresh cups since the 1970s—the variety of freshly made foods has exploded in recent years. Nachos, French fries, cotton candy, and pizza are all available, piping hot and fresh, at the press of a button. While it might be some time before vending machines produce fresh food that's not loaded with sugar or trans fats, the capabilities of vending machines seem almost endless.

Let's look closer at the cotton candy vending machine. You may be thinking that you put in your money and out comes a pre-made package of the fluffy treat. But you'd be wrong. In these modern machines, you put in your money and a fresh cone of cotton candy is spun before your eyes. Operating completely by sensors, the cone is first picked up and then inserted into the spinning bowl. The sugar feed automatically starts in the bowl, making sugar floss (see Chap. 10) that's collected on the cone. After a pre-set amount of cotton candy is spun, the arm retracts and drops the loaded cone into the chute for the purchaser to remove. It only takes about a minute or two to complete the process.

But back in my office, my options are limited to a few candy bars. Finding the perfect bar to get through the day can't be taken lightly. I scan the rows of candy one last time before finding the perfect energy boosting candy: Snickers. The combination of the instant sugar rush of chocolate and nougat with the staying power of peanuts makes it the ultimate midafternoon pick-me-up. I head back to my desk satisfied.

60

Snickers Bars

Apparently, Snickers bars are good for more than a sweet treat and quick energy. They're also good for demonstrating plate tectonics to grade school kids.

Imagine the Snickers bar in your hand is the Earth. Gently pull the Snickers bar apart and you observe a normal fault. This is where the "plates", or chocolate layer, separates to allow the Earth's mantle, or the caramel layer, to come into view. To observe a transform fault, push one half away from you and pull the other towards you. This shifts the plates horizontally relative to each other. And then push the two ends together to illustrate a thrust fault, where one "plate" goes under the other.

Snickers bars are a cool way to teach Earth science. And when you're done, you can eat your work.

Snickers bars have been a favorite for a long time. Named after a favorite horse, the Snickers bar was first offered by Mars in 1930 and quickly became a huge hit. A layer of vanilla nougat, topped with a layer of caramel and peanuts, and enrobed in chocolate provides a sweet treat.

Besides teaching Earth science, Snickers bars also teach us about effective marketing strategies. The campaign, "You're not you when you're hungry," is both entertaining and extremely effective. My favorite features Roseanne Barr as a hungry lumberjack complaining that her back hurts who then gets hit by a log and says, "now my front hurts." This long-standing campaign has featured numerous famous actors in similar roles, all of which are effective at getting across the message that a Snickers bar is a good option when you're low on energy.

R.W. Hartel and AK. Hartel, *Candy Bites*, DOI 10.1007/978-1-4614-9383-9_60,
© Springer Science+Business Media New York 2014

Snickers bars are also often used to demonstrate a common industry practice these days—cost reduction. Although Snickers bring in huge profits (about $3.5 billion dollars globally in 2012), Mars, like all large companies, continually tries to produce the same product day in and day out, but at lower costs. To enhance profits and offset rising ingredient costs, companies have a couple choices.

They could simply reduce the size of the bar sold at the same price. Mars apparently did that recently with Snickers, decreasing weight from 58.7 to 52.7 grams with calorie count decreasing from 280 to 250. With rising costs for almost every ingredient, including sugar, milk, peanuts, and chocolate, companies absolutely have to find ways to offset these costs and selling a little less to consumers is always one approach.

The second approach is called "cost reduction". Cost reduction is another way of saying let's make it cheaper. Finding ways to save costs (energy, time, etc.) in the manufacturing process or finding cheaper ingredients are both common cost reduction approaches used by companies (and not just food companies).

Mars hires food engineers to evaluate every step of every process for the products they make, including Snickers. Is there a way to cut some time off a step in the process or even to modify the process slightly to decrease costs without sacrificing quality? Are there new technologies for accomplishing each step more efficiently? These engineers spend their days sorting out different options to shave even a fraction of a cent off each product.

Food scientists are also involved in finding ways to reduce ingredient costs, again without sacrificing quality. Is there a cheaper source of milk for the caramel? Is there a cheaper source of peanuts or cocoa beans? The danger of this approach is that the quality of the product suffers. High end candy makers tout the quality of the ingredients as the reason their products are so good. And that's absolutely right, but it often requires passing higher ingredient costs off to the consumer.

To reduce costs, you can, for example, make caramel with dried milk powder, but it generally doesn't have the same appeal as caramel made with fresh cream. Even further, we can replace the milk fat in caramel with a cheaper vegetable fat and then add a butter flavor to offset the loss of flavor with the cheaper fat. It doesn't taste as good, but it saves money. Saving even a fraction of a cent on ingredient costs for a Snickers bar can save the company millions of dollars with the huge volume of Snickers sales.

How do you know you haven't sacrificed quality with a cost reduction move? Food companies use trained sensory panels to compare the cost reduced version with the previous version. If few people can detect the difference, the cost reduction was successful and the new model is launched.

The problem is that over a period of ten years of cost reductions, with each new iteration being tested against the most recent iteration, the quality of the product can significantly decrease. Even though each year's version may not result in a significant decrease in quality, the sum of ten years of cost reduction changes may profoundly affect consumer satisfaction. What used to be a delicious product may eventually become a product no longer desired by the consumer.

We can probably all remember products we ate as kids that no longer taste like we remember them. Perhaps part of that is just that we're growing up and either are more aware of what's in the food we eat or our tastes change, but some of it is the cost reduction mentality. For example, Coca Cola went from using sucrose as a sweetener to using high fructose corn syrup to reduce costs. Did product quality suffer from that change? Some say yes, and the recent proliferation of sucrose-sweetened soft drinks suggests that many consumers are demanding the original version. Still, from a business sense, it must have made cents for Coke, and all the other soft drink manufacturers, to make the change.

Companies that recognize this long-term quality effect from continued cost reduction have often reformulated again to make sure quality wasn't being sacrificed. Whether or not this story is completely pertinent to the Snickers bar, we can't be certain, but in principle cost reduction is practiced throughout the industry and has undoubtedly touched even this icon of popular candy. As in all modern candies, cost reduction is a reality and a necessity. But the aim is to reduce costs without sacrificing quality.

Besides being the number one candy bar in the United States, especially in vending machines (see Chap. 59), Snickers bars provide fodder for various lessons, from Earth Science to marketing strategies to the economics of cost reduction.

61

Baby Ruth

In the famous pool scene in the movie Caddyshack, a Baby Ruth candy bar gets thrown into the pool. To Jaws-like music, it floats away among the misbehaving kids. One by one, the kids see it and jump out of the pool, mistaking it for something else of the same color and general shape. Finally, the last kid in the pool swims up to it with a snorkel outfit, sees it, and yells "Doodie".

The movie cuts to the bottom of the pool after it's been drained. Bill Murray has a squeegee and finds the missing Baby Ruth bar. He picks it up, looks at it, smells it, and, in his own wacko style, takes a bite. "It's no big deal", he says, as if it was as good as the moment it came out of the wrapper. The old lady faints, thinking he ate—well, you know.

The joke works well in the movie, but how realistic is it? Can a chocolate-covered candy bar hold up to chlorinated pool water? Or will the water soak into it and make it a mushy mess? And, more importantly, is the density of a Baby Ruth bar, chocolate-coated nougat, caramel and nuts, less than that of water?

The chocolate coating on a Baby Ruth bar contains about 32 percent fat, and since fat and water don't mix, perhaps it would be completely impervious to water. However, chocolate also contains almost 50 percent sugar, in the form of numerous small crystals. Sugar crystals love water, as evidenced by the huge amount of sugar that can dissolve in water. Did you know that about 210 grams of granulated sugar can be dissolved into 100 grams of water at room temperature? And the warmer the water, like in a heated pool, the more sugar it can hold.

R.W. Hartel and AK. Hartel, *Candy Bites*, DOI 10.1007/978-1-4614-9383-9_61,
© Springer Science+Business Media New York 2014

What happens when a Baby Ruth bar, or any chocolate bar, comes into contact with water? Does it hold up to chlorinated pool water and remain edible for very long? In true scientific fashion, we set up an experiment to test this question. Several candy bars were immersed in standard, swum-in-every-day, run-of-the-mill pool water taken from the Northland College pool and observed hourly. You might be surprised at what we found.

First, let's look closer at a Baby Ruth bar. The Curtiss Candy Company in Illinois is credited with development of the Baby Ruth bar in 1916, a refinement of the Kandy Kake, an earlier product. The same company brought out the Butterfinger bar a few years later.

Numerous ideas have been floated (like the Baby Ruth bar in the pool water?) about how the Baby Ruth bar got its name. The company itself claimed that the bar was named after Ruth Cleveland, the daughter of President Grover Cleveland. However, she died 16 years before the candy was developed, making that origin a little shaky. Speculation is that the candy bar was named to take advantage of the fame of the Yankee baseball great, Babe Ruth. However, there's no evidence of this either. I guess we'll never know for sure how the Baby Ruth bar got its name.

Another fact of interest about the current Baby Ruth bar is the chocolate coating. It's not chocolate. It's a compound coating made to look, feel and taste like real chocolate, but it's made with palm kernel and coconut oils rather than cocoa butter. Nowhere on the label does it say chocolate—it can't because of the Standard of Identity that protects the use of real chocolate. The Baby Ruth bar is "bursting with peanuts, rich caramel and chewy nougat", but it contains no chocolate.

How does this affect its ability to survive chlorinated pool water? We'll see.

Chocolate is often used as a moisture barrier. Put a layer of chocolate between a layer of caramel and a cookie, and the water stays in the caramel, at least for a while. Without that layer of chocolate, the water would quickly migrate from the caramel to the cookie, resulting in hard caramel and soggy cookie. The chocolate slows down water migration but doesn't completely prevent it.

Chocolate, despite the high fat content, is not a perfect water barrier. In fact, in another experiment, we measured the change in water content of each part of a Twix bar, a chocolate-coated, caramel-covered cookie, and found to our surprise that the water content of both the caramel and cookie decreased with storage. Where did that water go? Out through the chocolate!

Because of the numerous sugar crystals in chocolate (and chocolate coatings), there is a tortuous path that water can follow through the chocolate. In our Twix bar experiment, the water from both the caramel and cookie slowly migrated out through the chocolate layer, causing the caramel to get hard. And this happened with no apparent change to the chocolate.

Back to our chocolate bars in chlorinated pool water. We put six different candy bars (Baby Ruth, Snickers, Milky Way, Twix, 3 Musketeers, and Tootsie Roll) into the water. First, the Baby Ruth bar, with peanuts, caramel and nougat, was denser than water and settled right down to the bottom of the pool. So right away you know the movie makers took some liberties with that scene. In fact, the only candy bar to float was the 3 Musketeers—not surprising, since it's just whipped up goodness coated in chocolate. The nougat is sufficiently aerated that the bar is less dense than water.

The candy bars also didn't fare so well sitting in pool water over night. The chocolate on all the candy bars, except for the top surface of 3 Musketeers that was out of the water, quickly turned a milky white as the sugar was extracted and dissolved. After 24 hours, the Baby Ruth bar had numerous cracks in the chocolate coating and the water was starting to dissolve the nougat and caramel center. What was left of the chocolate coating had no integrity and simply mushed off the candy bar when picked up.

There's no way that Bill Murray could have found a Baby Ruth bar that still was edible after spending a night in the pool. Although it was a funny scene in Caddyshack, Hazmat suit and all, there's no way a Baby Ruth bar would have floated and there's no way Bill Murray would have enjoyed a Baby Ruth bar after even a few hours in the pool. Rack that scene up to Hollywood invention.

62

Sometimes You Feel Like a Nut

It seems that coconut is one of those ingredients you either love or hate. We're split on it ourselves, with one of us loving it enough to make Mounds/Almond Joy a favorite candy and the other despising it. Why do some people so vehemently hate coconut?

Coconut grows on trees in tropic and sub-tropic environments. We've all seen cartoons where the coconut falls from the tree onto the head of an unsuspecting passerby. It's not actually a nut, as the name suggests, but really a fruit, or more precisely, a drupe. It has numerous uses, from making candies to the characteristic coconut aroma of sunscreen. Although its use in sunscreen may very well be to invoke the image of the tropics, it turns out that coconut oil is a good skin moisturizer. It's made mostly of lauric acid (a 12 carbon saturated fatty acid), a relatively short-chain fat that penetrates the skin more readily than longer-chain fatty acid chains.

When you think of coconut in candy, you probably think first of Mounds and Almond Joy. No wonder, these are the predominant coconut candy brands, although as we'll see later, far from the only ones. They've also been around a good long time.

The Mounds candy bar was developed by the Peter Paul Candy Company in the early 1920s. Who were Peter and Paul, brothers who went into business together? No. The company was started by an Armenian immigrant named Peter Paul Halajian (hence, the company name) and became one of the most widely known candy companies in the mid-1900s. The Almond Joy bar came out in 1946 and rivaled the Mounds bar in popularity. Their approach to marketing is unique—they sell two candy bars at the same time. "Sometimes you feel like a nut. Sometimes you don't. Almond Joy

R.W. Hartel and AK. Hartel, *Candy Bites*, DOI 10.1007/978-1-4614-9383-9_62, © Springer Science+Business Media New York 2014

has nuts. Mounds don't." Are there any other products that use that marketing approach?

Coconut in candy is usually in the form of dried, also called desiccated, shreds that can either be sweetened or unsweetened. The coconut "meat" is separated from the rest of the coconut and sent to a shredder to cut it into those stringy flakes that get stuck in your teeth, but that provide valuable fiber for your health. Commercial brands of desiccated shredded coconut contain added sugar to provide sweetness, propylene glycol as a humectant to keep it from drying out too fast, and sodium metabisulfite, a preservative. Between the two additives, they help to prevent that bag of coconut growing a bag of mold. While it would be more natural to forego the preservatives, that's not necessarily a good thing in this case. Shredded coconut meat quickly gets moldy—it's a great place for microorganisms to grow.

Although Mounds and Almond Joy are probably the first coconut candies you think of, there are actually many more candies that use coconut in one form or another. One old-time favorite that has now hit the deceased candy pile is, or was, the Brach's Neapolitan Coconut Sundaes. These candies, made of vanilla, strawberry and chocolate layers of coconut candy wrapped in a clear plastic wrapper, were found among the variety of candies in the, also now defunct, Brach's Pick-a-Mix. Still available is the Coconut Slice, another neapolitan-flavored, coconut-based confection.

Coconut Long Boys are a coconut-flavored caramel sold as a long stick candy. One blog describes their flavor as a cross between a Sugar Baby and the Coconut Neapolitans with a soft and nonsticky texture. Then there are chocolate coconut haystacks, simply chocolate solidified with coconut shreds spread throughout. Coconut admirers love that mixture of sweet coconut and chocolate, while coconut detesters wouldn't go near it.

Why is coconut one of those things that you either love or hate? One blogger claims that "it's cloying, sickly sweetness and chewy-yet-flaky texture" are what sets her off. Most people don't like it for that chewy consistency. But they're only speaking of the desiccated shred product, like what's used in the products mentioned above. Coconut can be so much more.

For example, toasted coconut provides a unique and delectable flavor for confections. The Zagnut, developed in 1930 by the Clark Company (also renowned for the Clark Bar), is a crunchy peanut butter and toasted coconut bar. Chick-o-Sticks are another crunchy peanut butter and toasted coconut candy. One candy, Marshmallow Coconut Toasties, made by Melster Candies (now Impact Confections) in Wisconsin, combines the airiness of a marshmallow with the flavor of toasted coconut.

The Maillard browning reaction (see Chap. 28) between sugars and proteins and caramelization of sugars both contribute to the unique toasted coconut flavor. Coconut meat contains sufficient sugars to react with the proteins present, although sometimes additional sugars are added during toasting to create different flavors. Toasting coconut also generates flavor reactions from the coconut oil (in the same way that cooked butter flavors contribute to the caramel flavor).

Perhaps one reason some people, especially older people, don't like coconut has to do with the accusation many decades ago that coconut oil (and other so-called tropical oils) promotes heart disease due to the high saturated fat levels. This has proven untrue and, in fact, coconut has numerous health benefits over and above the benefits of skin health. For one, coconut oil contains medium chain triglycerides, which are fats that could potentially help with weight loss, promote immune system health, and even promote heart health. Further, coconut meat has high fiber content, something most of us don't get enough of. Eating and drinking coconut-based products is now actually considered to be good for your health.

Further, the recent popularity of coconut water as a healthy drink attests to the change in public perception and increasing interest in coconut. Touted as a natural sport drink, coconut water is an excellent source of potassium (more than four bananas!), along with sugars and electrolytes to enhance hydration. It may not cure cancer or ease a hangover, as some marketers claim, but it is a proven hydration drink, with fewer calories than products like Gatorade.

To promote coconut consumption, companies have been developing new and unique forms that are significantly better than the desiccated shreds that so many people detest. Coconut chips and chiplets, whether sweetened or unsweetened, toasted or untoasted, provide unique options for creating delicious and nutritious coconut products, including confections.

If you're one who claims to hate coconut because all you can think of is the desiccated shred, then run to the store and look for the latest and greatest in coconut ingredients and products.

63

Turtles or Cow Pies?

Of all Aesop's fables, arguably his most well-known is the story of the Tortoise and the Hare. The moral of this classic fable is that the hard-working, steady consistency of the tortoise wins the day over the fast but flighty hare. A candy version of Aesop's Tortoise and the Hare might be the Turtle and the hollow chocolate rabbit.

A Turtle is a well-loved candy, caramel drizzled on pecans and topped with chocolate. A delectable combination of sweet and salty, these candies get their name from their appearance. Pecan legs and head peek out from underneath a chocolate-covered caramel shell. You can almost envision it lining up for Aesop's race alongside the hollow chocolate rabbit.

Be careful what you call a Turtle though. Like Band-Aids are technically adhesive strips, most turtle candies should probably be called something like chocolate-covered caramel pecan clusters. The word Turtle is a registered trademark (and should technically always be written with the ® symbol), for one specific brand of turtle candy.

The history of the turtle candy is a little cloudy, with several claims being made on the origin, including one by the current trademark owner, DeMet's Candy Company, in the United States. Technically, they're the only company in America that can market this treat as a Turtle (Nestle owns the trademark internationally). Other companies must get more creative.

The traditional way to make turtle candies involves pouring a mound of fluid caramel onto a layer of pecans, or other nut, on a tray or marble slab. Shake off the nuts that weren't attached to caramel and then pour a layer of tempered chocolate on top of the

R.W. Hartel and AK. Hartel, *Candy Bites*, DOI 10.1007/978-1-4614-9383-9_63,
© Springer Science+Business Media New York 2014

caramel. If done right, the mounded dome of chocolate-coated caramel will have pecans sticking out that may (or may not) look like feet and head.

The process has been automated so that turtles can be made as fast as hares at breakneck speed on high throughput manufacturing lines. As with most automated processes, though, the individuality of handmade goods is lost; they all look exactly the same. Rows and rows of identical turtles now race off the conveyor at breakneck speed. That uniformity isn't absolutely necessary, though, since the Goo Goo Cluster, a cousin of the Turtle, actually has variability designed into its high speed process line (see Chap. 66).

The process must be sequenced and staged correctly to ensure a high quality turtle, whether made at home or in the factory. The caramel has to be soft and gooey when it contacts the nuts in order for them to stick, but then the caramel needs to be firm enough to hold the chocolate application. At the end, the entire piece has to be held at the correct temperature to allow the chocolate to set up. In the factory, a naked turtle (without it's chocolate shell yet) passes through an enrober to get its chocolate coating and then enters a long cooling tunnel to set the chocolate. When it exits, it's sufficiently solidified to be ready for the package.

A similar type of candy can be made with other nuts like cashews and peanuts and with any type of chocolate (dark, milk or white). And this helps get around the trademark restriction.

There are numerous creative names (and descriptions) that candy makers use to get around the trademark. Fanny May Candies makes Pixies—pecans, rich caramel and luscious real chocolate. See's Candies, who some consider to be the original source of the turtle candy, markets Polar Bear Paws, a peanut version of the turtle made with buttery caramel and white chocolate. An old-time candy version of the turtle was called the Katydid, from Kathryn Beich candies. It used to be sold in a keywind can (like old cans of Spam or sardines). To get to your Katydid, you had to insert the key, usually found attached to the bottom of the can, into the slot of a band that ran around the can. Winding the key opened the can.

Although more difficult to access, the can provides an outstanding package for extended shelf life.

One candy company calls turtle candies Chocolate Tortoises, actually evoking Aesop's fable on their web site to note the patience and steady attention needed to create their artisanal caramel. Choosing tortoise also helps them get around the trademark issue.

Although the tortoise is often replaced by a turtle in the fable (sort of like we did in this chapter), that's technically incorrect. Both are reptiles from the same family and have shells into which they can withdraw, but those shells are significantly different based on their primary habitats. Since the tortoise dwells on land, its shell is heavier and boxier than the turtles. Turtles live primarily in water so their shells are lighter, flatter and more streamlined.

In truth, the fable couldn't be about a turtle and a hare racing on land. They live in different domains. That would be akin to a porpoise racing a cow; whether they raced on land or in water, one would be out of its element.

Speaking of cows, in Wisconsin we have our own version of the turtle—the Cow Pie, from Baraboo Candy Co. I guess where most people see a turtle, someone from cow country sees a cow pie (careful, a cow pie throwing content might break out at any moment). The turtle-like Cow Pie contains "fresh pecans and gourmet signature caramel smothered in rich milk chocolate."

Some creative line extensions from Baraboo Candy include the Peanut Butter Cow Pie, fresh roasted peanuts covered with a layer of signature caramel, a layer of creamy peanut butter and smothered in rich milk chocolate and the Green Bay Puddle, fresh peanuts, creamy caramel and rich milk chocolate. Homer's Snowflakes are another variation, with fresh cashews, creamy caramel and rich white coating.

Whatever you call them, chocolate-coated caramel nut clusters are a delightful treat.

64

Candies: Dead or Alive

Chicken Dinner Candy Bar? Tween Meals? 3 Pigs? Dr. I.Q.? Candy Dogs? Cold Turkey? Dick Tracy Candy Bar? Reggie Bar? Denver Sandwich Bar? High Noon Candy Bar?

All candy bars that have come and gone. Seen their best day. Only memories, at best.

Why have some candies developed into big hits and others faded into oblivion (or the old candy wrapper web sites)? Maybe the Snickers Bar just tastes better than the Chicken Dinner Candy Bar?

The Chicken Dinner Candy Bar was "a nut roll covered in chocolate". What's not to like? It was a big hit in Milwaukee, WI for 50 years, from the 1920s to the 1970s, but then the company was sold, and then sold again. Even with a fleet of Chicken Dinner trucks (picture an Oscar Mayer Wienermobile but shaped like a chicken instead of a hot dog), marketing couldn't save it.

Even Mars had some interesting bars that didn't make it big. The Dr. I.Q. bar? From the old wrapper, it was "a delicious treat of nougat and smooth rich caramel, roasted peanuts, all covered with the finest milk chocolate". Seems like it was pretty close to the Snickers Bar.

Maybe it's in the name? Dr. I.Q.? Snickers? It's hard to imagine a Dr. I.Q. bar taking off like Snickers did.

For whatever reason, sometimes candy brands die out. It may simply be that tastes change, sales decrease, and a company decides that a brand no longer makes enough profit to warrant its continuation. No sense in making a candy that no one eats, or more to the

R.W. Hartel and AK. Hartel, *Candy Bites*, DOI 10.1007/978-1-4614-9383-9_64,
© Springer Science+Business Media New York 2014

point, buys. No sales, no profits, no candy. Maybe Bit-O-Licorice and the all-black licorice Chuckles fit into this category.

Sometimes the company wants to reposition a candy to take advantage of other more popular brands. For example, the Mars Bar was discontinued and brought back as a new product, Snickers Almond. Another example is the Forever Yours bar, renamed Milky Way Midnight. Both bars play on the branding of another, more popular, candy bar. Seems to make business sense.

Sometimes candy bars are introduced to recognize the accomplishments of a renowned sports figure; for example, Reggie Jackson, Mr. October. A professional baseball player, he was especially prominent in the post-season and once hit three consecutive home runs in a World Series game. For his efforts, the Reggie Bar was developed in his name in 1978. Although he wasn't really a one-hit wonder, the Reggie bar lasted only a few years while his star faded.

The Stark Candy Company provides a good example of what can happen to candies. Starting in 1937 outside of Milwaukee, WI, the Stark Candy Company was known for making such products as Sweethearts Conversation Hearts and Candy Raisins, a honey-colored Dot-shaped candy with a vaguely raisinish flavor. One remains a successful product while the other has been discontinued. When NECCO bought Stark in 1988, they continued to run the Wisconsin plant, making both products. Note that another Stark favorite, Snirkles, a caramel roll, had already been phased out by the time NECCO came in. When the business decision was made to close the old Stark plant, production of Sweetheart wafers went to another plant while production of Candy Raisins was discontinued. Why? Probably because conversation hearts were a national candy while Candy Raisins were just local.

There has been enough backlash about the discontinuation of Candy Raisins, though, that efforts have been made to bring them back. But since NECCO owns the recipe and secrets of processing, attempts to reproduce the product have been largely unsuccessful. Plus, NECCO still owns the trademark, meaning a new name is required and some of the appeal of the original is lost.

Another example of a discontinued candy product that has seen a recent comeback is the Pine Bros cough/throat drop. A hard, yet "softish", gummy candy made from gum arabic, Pine Bros. throat drops were developed in 1870 as a lozenge to relieve a sore throat. Although more medicine than candy, many kids claimed a sore throat in hopes of being "treated" with one (or more) of these drops. They had a long and successful run, but eventually the brand got traded one time too many and the business decision to drop the line was made.

However, driven by the popularity of retro candy, Pine Bros. throat drops are once again being produced, but by a completely different company. Here's how that happened.

First, someone with bucks decides to reintroduce the candy. She finds an old employee of the company with some experience with the product to reinvent the formulation. They then identify a company willing to help them produce the product, often called a co-packer in the food industry, who then hires a research lab (at the University of Wisconsin) that can help develop the process and perfect the flavor. Finally, if everything goes well, Pine Bros cough drops that are similar to the original, or at least someone's memory of the original, are out on the market. Assuming there is enough retro demand and sufficient marketing to create a new generation of users, the product once again becomes a successful money maker.

Another example of a successful reintroduction of a deceased candy is Bonomo's Turkish Taffy. I have fond memories of Bonomo's as a kid growing up on Long Island. Even the old commercial was great. B-O ... N-O ... M-O ... Oh, Oh, Oh It's Bonomo's ... Caaandy! Freeze it, smack it on the counter and eat the bits. It's been back now for several years and appears to be successful. Even though it would be murder on my teeth if I were to eat it now, it's good to see it come back.

Yet, the process of candy rejuvenation doesn't always work that well. For example, in the process to bring back a Candy Raisin knock-off, a co-packer candy company hasn't yet been identified so the person with the big bucks has to decide whether or not to invest the money himself in a manufacturing line. Even though he's got a good flavor match to the original, it's risky because there's no guarantee the demand will be sufficient to make this a successful venture. And in the meantime, another company is trying to market a different version of Candy Raisins, although so far with marginal success at matching the flavor and texture of the original.

The graveyard of deceased candies is a fun place to look for nostalgic brands and for interesting, sometimes wacky (a la Chicken Dinner bar), old candies. Which candy bars would you phase out? While all candies have their fans, candy making is a business and only products that have a large enough following warrant continued production.

65

Super-Sized Candies

In the past few decades, candy companies have been promoting the smaller versions of original products, but sometimes there's also the super-sized version to choose from. Minis and fun size candies provide options for eating less than the original-sized bars, but recently there's been a growing market (pun intended) in the other direction, with many companies producing larger-than-life products. From giant gummy bears to a one-pound Snickers bar, these super-sized products cater to a specific niche.

The one-pound Snickers bar, called Slice 'n Share, has been on the market for a few years and, according to a company representative, is seeing increasing sales. With over 2,000 calories, this giant candy bar is definitely not a single serving size, unless you want to get your entire day's calorie quota at one go. It's promoted as a fun product to share—slice and share. We stopped in at the plant where they make the Slice 'n Share last year; although they weren't running Snickers at the time, we got the low down.

It's so big that it causes manufacturing problems. It can't be run on the usual machines that make Snickers bars so it has to be made by hand, like in the old days. Candy makers create separate batches of nougat and caramel on small-size equipment. They then assemble the candy bar by hand, carefully layering the caramel on top of a layer of nougat and sprinkling the peanuts as needed. The giant candy bar is then enrobed in chocolate, cooled and packaged by hand. I suppose if the demand for this behemoth were big enough, the engineers would develop a process that could handle the larger sized bars. Until then, they'll be hand made.

R.W. Hartel and AK. Hartel, *Candy Bites*, DOI 10.1007/978-1-4614-9383-9_65,

All of the super-sized candies have a similar problem—they're too large to be made on commercial process equipment and must be made by hand. And often with fairly unique considerations because of the extra size.

How about the one-pound Sugar Daddy? More like a paddle than a candy piece, this huge Sugar Daddy also has to be made individually. At 4.5 inches wide and 17 inches long, it has to be carefully constructed. Again, no automation is possible. The liquid caramel is poured into a specially-made mold, the stick is inserted in the side, and the piece is allowed to cool, after which it's removed by hand and packaged. Because of the manpower required to make each one, production is limited.

It's the same for those enormous lollipops. The all-day sucker is made by pouring the molten sugar candy into specially-designed molds. They're then allowed to cool and solidify before being popped out and packaged.

Years ago, we found a one-pound marshmallow heart. Since a typical marshmallow is more than half air, it takes a lot of space to make a one-pound, sugar-sanded marshmallow. It's really in the shape of a heart—not a valentine heart, but an anatomical heart, with blue blood vessels painted on the surface and all. At one-pound, it's definitely larger-than-life. From its shape, it had to be deposited, undoubtedly made by hand. That is, the warm marshmallow candy mass was poured into a special mold, whether made of plastic or corn starch we don't know, where it solidified as it cooled. The mold was then broken off to reveal the heart shape with the different chambers of the heart and openings where the arteries come out. It must have been sugar-sanded by hand as well before being packaged. It's definitely a pretty unique addition to the enormous candy collection; unfortunately, they're no longer being made. I guess the market wasn't large enough to bear the cost. It will remain a museum piece on my candy table.

Another interesting super-sized candy is the huge jaw breaker, more than a mouthful for sure. The Mega Bruiser, weighing a pound with a wing span over three inches, takes several weeks to make. Jaw breakers are made by continually building a sugar shell on the piece while they're rotating in a large pan (see Chap. 46). Imagine softball-sized candies weighing nearly a pound tumbling in a rotating drum—the noise has to be deafening and the force with which each candy hits the drum wall must be enormous. In fact, pans that run these monsters have to be made from tough stainless steel, not copper, to withstand the mega-Newtons of force.

How do you eat one of these huge candy balls? They're way too big for the average mouth. Simple, hit it with a hammer until it's in bite-sized pieces. Just like they do with the hard candy Peppermint Pig from Saratoga Springs. Except the pig comes with its own hammer.

Other large candies include Mega Smarties, the Giant Pixy Styx, and Big Tex Giant Jelly Beans. These jelly beans are over an inch and a half long and nearly an inch wide. One of them is definitely a mouthful. In fact, serving size is supposedly one bean—worth 45 calories (zero from fat). Unfortunately, the candy company in Texas that was making Big Tex has apparently closed the plant due to the high sugar costs. Whether we'll be seeing Big Tex jelly beans in the future is as uncertain as the sugar subsidies (see Chap. 5). With production potentially moving to Mexico, maybe they'll be renamed Big Tex-Mex jelly beans.

Gummy bears are normally pretty small. There are about 175–200 regular-sized pieces per pound, depending on the brand. That makes the five-pound gummy bear, mistakenly called the "world's largest" by one site, the equivalent of close to a thousand individual gummy bears. Five pounds is enormous, but not nearly the largest. One company regularly makes seven-pound gummy bears as give-aways to sales brokers.

And then there's the 26-pound party gummy bear, claiming to be over 5,000 times larger than a regular gummy bear. What's unique about this one is its bear (not beer) belly. That's large enough to hold 34 ounces of candy or liquid. Lay this candy bear down on its back and use the bear belly as either a candy bowl or a punch bowl at your next party. Once the innards are gone, slice up the bear to finish off the treat.

66

Goo Goo Clusters

Think of yourself as a candy bar aficionado? Do you think you would be able to distinguish which candy bar is which just from looking at a cross-section? If so, challenge yourself visually at the Science Museum of Minnesota's Thinking Fountain web site, where you can see numerous candy bar cross-sections to see if you can guess what they are.

For now, here are a few verbal cross-sections for you. A layer of vanilla nougat followed by a layer of caramel with peanuts, all coated in milk chocolate? Snickers of course. Wafers with chocolate cream, coated in milk chocolate. Easy, Kit Kat. A layer of chocolate nougat topped by a layer of creamy caramel, all coated in milk chocolate? Right, Milky Way. How about a nougat center surrounded by caramel with peanuts all around, coated in milk chocolatey coating? Not so easy? The answer's down below, but read on before you look it up.

One more—a disk-shaped candy with a layer of marshmallow nougat topped by a layer of caramel, covered with peanuts in milk chocolate? Of course, it's obvious from the title of the chapter. It's a Goo Goo Cluster.

Goo Goo Clusters were developed in 1912 by the Standard Candy Company and are still made today in Nashville, TN. Some speculation exists about the name, but according to the Standard Candy web site, the name originates from a suggestion by a fellow bus rider to the inventor that it's so good, people will ask for it from birth. Hence, it's named after the first words a baby says.

Goo Goo Clusters claim to be the first combination candy bar. Nougat, caramel and chocolate were all well known before 1912,

R.W. Hartel and AK. Hartel, *Candy Bites*, DOI 10.1007/978-1-4614-9383-9_66, © Springer Science+Business Media New York 2014

but no one had thought to put them together in one bar before. Exactly what they were thinking when they came up with the idea to make layers of the different candies and coat them all in chocolate? Who knows, but their candy bar baby was an important step forward in confectionery development. Shortly after, several more combination candies, or candy bars as they've come to be known, came out. The Goo Goo Cluster was quickly followed by the Heath Bar and O'Henry.

The first Goo Goo Clusters were undoubtedly made by hand; first a layer of nougat was spread out, followed by a layer of caramel spread on top, and then they were coated in peanuts and milk chocolate. From careful inspection, it appears that there are two layers of chocolate coating application between which a sprinkling of peanuts is applied. The first layer of chocolate allows the peanuts to adhere before getting the second coating. Each unit was probably a bit unique as the candy maker created each one individually.

These days, Goo Goo Clusters are made on an automated line, at a rate of 20,000 per hour. That's a lot of candy. Has automation changed their appearance? Not so you could tell. We cut a couple of them up recently to look at them, and they all look slightly different, as if they were still made by hand. Usually, when a production line is automated, everything is carefully controlled to be the same all the time, including the appearance. When Standard Candy automated the Goo Goo Cluster line, they must have built in some variability to make sure each piece looks slightly different. Other attributes, like the exact weight of each piece, are undoubtedly controlled very carefully, but the appearance seems to have been carefully "uncontrolled" to retain that hand-made look. Perhaps the random orientation of the peanuts as they sprinkle onto the first chocolate coating layer is sufficient to provide random shapes.

Although Goo Goo Clusters were first, they aren't the only candy in the Goo Goo line-up for Standard Candy. Since the early 1980s, you can also get a Goo Goo Supreme, made with the classier pecan instead of the pedestrian peanut. And for those who love peanut butter and chocolate, there's a Goo Goo Peanut Butter, introduced in 1991. It's a layer of peanut butter covered in peanut-laden milk chocolate.

What exactly is marshmallow nougat, the base layer of a Goo Goo? Goo-google marshmallow nougat and not much comes up, primarily because they're typically considered different things. Although both are aerated candies, marshmallows are usually much more aerated than nougat (which is correspondingly more dense). However, often nougat is made using frappé, a type of marshmallow. Frappé is made by whipping sugar syrup with a protein stabilizer to make a whip (rather than using gelatin like a good chewy marshmallow). Goo Goo Clusters contain several proteins in the ingredient list, including milk proteins and soy proteins.

Another interesting thing about Goo-Goo Clusters is that the last few ingredients on the label are all nuts—almonds, pecans, walnuts, cashews and hazelnuts. Why are they there? It could be that they're added at low levels, even less than the added salt, to jazz up the label. Nuts are good for you, right, so why not add some of each? But their reasoning is different—they do it because it allows them to have a clean allergen label. In reality, the Goo Goo Cluster does contain these tree nuts. Nothing ambiguous like "may contain nuts." They do contain those nuts.

The answer to the last verbal quiz question from above sounds like it could be the Goo Goo Cluster, but it's not, it's Baby Ruth.

One last quiz for you candy bar aficionados. An interior layer of soft chewy caramel, coated in milk chocolate laced with crisp rice? Why, a 100 Grand bar of course. As their slogan says, "That's Rich."

67

Candy Land*

There's a persistent story that percolates through our family's mythology. Like mythology, the details have become hazy, so hazy that no one can actually remember when it happened or if it ever happened at all. But it's one of those stories that gets pulled out at parties or with family friends, used as an example that shows an underlying truth about my character or to commiserate with anyone that's raised a young child.

The story goes that as a toddler, I was so upset that I was losing at Candy Land that I threw a tantrum so epic, it's reverberations are still felt today. This tantrum was so bad that my mother banished all competitive games from our house for years. Any games that managed to slip into the house were played solo, against imaginary opponents. It was the only way to ensure no one got hurt.

Part of the sticking power of this story is that Candy Land is not a game most people would get upset about. Players follow a multicolored path as it winds through different candy themed locations by drawing cards at random. And that's it. For anyone older than four, it's mind-numbing tedium. Statistically-inclined parents have even published papers trying to find the fastest way to finish the game.

But to a particular kind of three year old, getting to King Kandy's palace first meant everything.

Candy Land was invented in the last year of World War II when Eleanor Abbott was laid up recovering from polio. I can imagine her staring out of her bedroom window, watching neighbors and

---

*Primarily by AnnaKate Hartel

R.W. Hartel and AK. Hartel, *Candy Bites*, DOI 10.1007/978-1-4614-9383-9_67,
© Springer Science+Business Media New York 2014

friends playing on the sun-soaked pavement of San Diego. Maybe they were using chalk to play hopscotch. I see her calling for paper and crayons, drawing her own little version, adding characters based around the sweet treats she was desperately craving.

In reality, Eleanor Abbott was a polio-stricken schoolteacher in San Diego; she invented the game to entertain the children in the hospital. The loopy and uncomplicated game was meant to distract even the youngest child without needing much adult supervision. Different locations where players were trapped for a turn (Molasses Swamp, Cherry Pit, and Gumdrop Mountain) helped to keep the game purposely slow, much to the later chagrin of parents worldwide.

The game was a hit in the polio ward and Abbott sold the idea to Milton Bradley (now called Hasbro) in 1949. Since then, the game has sold millions of copies and entertained both sick and healthy kids alike. While the specifics have changed throughout the years, the basic premise and game play has remained the same.

Sometimes I wonder if Ms. Abbott ever encountered the particular kind of three-year-old that I was, but I doubt it. It's more likely she came up against a more typical kind of three-year-old, the kind that throws a tantrum to get candy. While many of today's most popular candy bars were already well established by the mid twentieth century, I can't imagine they were plentiful in a hospital. For those kids, sweets were probably a very occasional treat.

Like Candy Land the game, candy itself is inextricably linked to childhood, a time when strong memories are being cemented into the growing brain. Enjoying a piece of candy can be like sensory time travel, conjuring up summers spent nibbling licorice in the park or days spent in the kitchen cooking up Grandma's famous fudge. These involuntary memories (a term first used by Proust in *Remembrance of Things Past*) are often triggered by some association with candy, like the aroma of fudge being cooked (see Chap. 1), because they are cemented into the brain during the development years. For kids, candy (or playing Candy Land) is an emotional experience.

You probably picked up this book because you have an emotional connection to one (or more) of the candies we've discussed. If that's the case, we hope you replayed those memories as you read those chapters. And we hope you can make a little bit of room for the science along with those memories.

24971055R00159

Made in the USA
Middletown, DE
13 October 2015